FRIENDLY BIOLOGY

Joey Hajda DVM MEd

Lisa B. Hajda MEd

Student Workbook

Published by Hideaway Ventures, 79372 Road 443, Broken Bow, NE 68822

For information regarding this publication please contact Joey or Lisa Hajda at the above address or visit our website: www.friendlybiology.com.

Front cover photo credits: *Tramea onusta.* (Red Saddlebags dragonfly) Tim Hajda; *Papaver spp.* (poppy) Joey Hajda; *Thamnophis spp.* (gartersnake) Tim Hajda and our granddaughter Phoebe Jean by her mother, Clara Williams, Clara Williams Photography.

Table of Contents

OUR PETS

TROTWOOD

BUD

SHIRLEY AND OLIVE

MARTHA, BABY 1, BABY 2, ETHEL AND GEORGE

VELVET AND TIM

JEAN LOUISE AND BONNIE

IF YOU HAVE PETS, DRAW THEM HERE!

Name ____Date_____________________

Lesson 1

Practice Page 1

Instructions: Fill in the blank with the appropriate word. Refer back to the text portion of the lesson for help.

1. The term biology is derived from two words: __________ which means life and ________ which means the study of.

2. There are other words like biology. The study of the earth is _________________ while the study of ancient artifacts is ______________________.

3. Hematology is the study of _________________________ and the study of tumors or cancers is known as ______________________.

4. Being able to ______________ or change location or position is evidence of life in a living thing.

5. A feature common to all living things is that they can make new living things. This process is known as ______________________.

6. Living things all require a source of __________ in order to carry out life processes. For humans this source is the ______________ we eat, however for most plants, it is the __________ which provides this source.

7. Small living things increase in size and complexity. Living things ______________ and develop.

8. Living things react or ___________________ to things around them which is also known as their environment.

9. _______________ is the condition where evidence of life no longer exists.

10. The name of the primary author of this textbook is ______________________.

Read each example below and tell whether which form of evidence of life is being demonstrated. Some examples may have more than one evidence of life present.

Example: A puppy chases a ball. Movement

11. Tommy is hungry and so he eats three hot dogs for lunch. ________________________

12. A tree sparrow flees from an approaching cat. ________________________

13. The goose in incubating ten eggs. ________________________

14. The earthworm wriggles down deeper into the bedding in the bait cup.

____

15. The sunflower tilts its flower towards the east in the morning hours.

____

16. Mushrooms tend to be found most frequently near dead or dyeing trees.

____

17. The kitten learns to catch mice by imitating its mother. _______________________

18. As the summer month go by, the ears of corn on the corn plants appear to be getting larger in diameter. _______________________

19. Allowing food to sit out unrefrigerated and then eating it can result in food poisoning.

____

20. The peach tree developed pink blossoms as the temperatures increased.

____

Choose an animal in your home or one you are familiar with. Tell what it is and then give four pieces of evidence that tells you it is a living thing. For example, if you choose your dog, a piece of acceptable evidence is that it needs to be fed everyday.

21. _______________________

1.

2.

3.

4.

Choose a plant in your yard or one you are familiar with. Tell what it is and then give four pieces of evidence that tells you it is a living thing.

22. _______________________

1.

2.

3.

4.

Name ____ Date______________________

Lesson 1 Practice Page 2

Instructions: Below you will find clues to solve this crossword puzzle. Refer back to the text portion of the lesson for help.

Across

2. living things _____ and develop
3. process of a living thing making another living thing
4. requirement of living things in order to carry out life processes
6. when evidence of life no longer is present in an organism
8. living things _____to their environment
10. author of this textbook
11. tumors or cancer
12. life

Down

1. changing position or location as indicator of being alive
5. earth
7. one complete living thing
9. the study of

Name ______________________________ Date ______________________

Lesson 2 Practice Page 1

Instructions: Fill in the blank with the appropriate word. Refer back to the text portion of the lesson for help.

1. All things, whether living or non-living consist of tiny bits of matter known as _________.

2. The central portion of an atom is known as the _________________ and it contains subatomic particles known as the ______________________ and the __________________.

3. Circling around the nucleus of the atom are a third type of subatomic particle which are the _________________________.

4. Theories say it is the ________________________ of the electrons which determines the behavior of various elements on the periodic table.

5. To determine the number of protons or electrons an atom of a particular atom has, one looks for the __________________________ of that element on a periodic table.

6. The electrons are thought to exist in __________________ around the nucleus of an atom and that there can be no more than ________________ on one of these layers.

7. Elements which have their outermost layer of electrons filled are the elements which are very _________________________ in their behavior.

8. Elements which have their outermost layers incompletely filled are elements which are very _________________________ in their behavior.

9. Elements which are very reactive seek to gain stability by moving or sharing _____________________ with neighboring atoms of elements.

10. The family of elements whose atoms have their outer layers of electrons completely filled making their very, very stable is the ____________________ family.

11. Atomic bonds which form between atoms who have transferred electrons from one to another are known as ______________ bonds.

12. Atomic bonds which form between atoms who are sharing electrons between themselves are known as __________________ bonds.

13. Of the many elements known to man, there are four that are common to all living things and deserve our special attention. Those four elements are: __________________, __________________, __________________ and __________________. Write their element symbols next to their names, too!

14. Of the elements listed below, choose the one that would most likely be the least reactive.

    A. Hydrogen

    B. Carbon

    C. Sodium

    D. Neon

15. Of the elements listed below, choose the one that would most likely be the most reactive.

    A. Neon

    B. Sodium

    C. Argon

    D. Helium

16. Which subatomic particle is thought to be responsible for an atom's behavior?

    A. Proton

    B. Neutron

    C. Electron

    D. Crouton

17. Suppose Atom A desires to get rid of one electron and Atom B is willing to accept that one electron. Together, by moving this electron, they can become a compound which is stable. This type of bond formation where electrons are moved is called a

    A. Proton bond

    B. Single covalent bond

    C. Double covalent bond

    D. Ionic bond

    E. James bond

18. There are many elements required by living things in order to maintain life. There are four that we will study next. Circle these four important elements found in this list of element symbols:

H He C Ca Li Be Ar Ox O Br Ni N Zn As Pb Cu

Name ________________________________ Date ______________

Lesson 2 Practice Page 2

Instructions: Below you will find clues to solve this crossword puzzle. Refer back to the text portion of the lesson for help.

Across

3. subatomic particle analogous to planets in our solar system
6. defined as a change in position or location, a feature of living creatures
10. It is the _______ of electrons which determines reactivity of an element.
11. a feature of living things where more living things just like the parents are created
13. an important element found in living organisms, a component of our air
14. the central part of an atom

Down

1. the type of bond formed when electrons get transferred from one atom to another
2. an important element of living organisms which has the atomic number six
4. tiny bits of matter from which all things are thought to be made
5. the type of bond formed when atoms share electrons between themselves to gain stability
7. an important element of living organisms which has the symbol N
8. an important element in living organisms, also found in water
9. electrons are arranged in ______ around the nucleus of an atom
12. subatomic particles found in the nucleus of atoms

Name ____Date_____________________

Lesson 3 Practice Page 1

Instructions: Fill in the blank with the appropriate word. Refer back to the text portion of the lesson for help.

1. The common name for carbohydrates is ______________________.

2. The word carbohydrate is derived from two words carbo– which indicates the presence of ______________________ and -hydrate which means ______________________ is also present.

3. The generic formula for a carbohydrate is ______________________ where the little letter "n" tells ____.

4. The carbohydrate in which n = 6 has the formula ______________________ and its name is ______________________. Its common name is ______________________,

5. Another carbohydrate in which n = 6 and is found in fruits is known as ______________________. Its structure is slightly different from that of glucose and is therefore called an ______________________ of glucose.

6. A carbohydrate made from one type of sugar molecule is known as a ______________________ where the prefix __________ means one and the root word ______________________ means sugar.

7. Two examples of a monosacharride are ______________________ and ______________________.

8. A carbohydrate which consists of two types of sugar molecules is known as a ______________________. A common example of this type of sugar is ______________________ commonly known as table sugar.

9. The two monosaccharides which make sucrose are ______________________ and ______________________,

10 ______________________ is the carbohydrate found in dairy products. It is classified as a ______________________ because it is made up of two simpler sugars. Those two are ______________________ and ______________________,

11. The fuel for living things is ______________________.

12. More complex carbohydrates can be broken down into glucose through the action of ______________________.

13. Enzymes primarily work in two ways: ______________________ or ______________________.

14. The scientific word for cutting is to ______________________.

15. The enzyme which breaks down table sugar is ______________________.

16. The enzyme which breaks down milk sugar is ______________________,

17. People who are deficient in producing enough lactase are said to be

____,

18. ______________________ is a disaccharide found in seeds and grains. The enzyme which breaks down this carbohydrate is ______________________,

19. Plants have the capability of joining several glucose units together into complex structures. These complex structures are known commonly as ______________________ and scientifically as

______________________,

20. The enzyme capable of lysing amylose is ______________________ and can be found in our

______________________,

21. ______________________ is a carbohydrate that is not sweet tasting, but instead functions to provide strength and support to plants. It is formed by very long chains of glucose molecules bonded together.

22. The enzyme which breaks down cellulose is known as ______________________.

23. Microorganisms within the stomachs of ruminants and termites work to the benefit of their host species. Tell or draw a picture of how this relationship happens.

24. A type of relationship where one organism gains benefit while the other suffers is known as a

______________________ relationship. Give two examples of this type of relationship.

1.

2.

Name ____Date____________________

Lesson 3

Practice Page 2

Instructions: Below you will find clues to solve this word find puzzle. Refer back to the text portion of the lesson for help. Note that the words may read forward, backward, up, down or at a diagonal.

1. Common name for carbohydrate is a ____________.
2. Cn(H2O)n is the generic formula for a ______________.
3. Carbohydrates are made up of three elements: ________________, ________________ and ________________.
4. Another name for a carbohydrate is a ______________.
5. A carbohydrate made up of only one type of sugar molecule.
6. A carbohydrate made up of two types of sugar molecules, such as sucrose.
7. These compounds can work like matchmakers to speed reactions or scissors to cut apart chemical compounds like carbohydrates.
8. Scientific name for table sugar is ________________.
9. ________________ is the carbohydrate which fuels living things.
10. This sugar makes the sweetness of fruits.
11. The carbohydrate of dairy products is ________________.
12. Plants store carbohydrates in the form of ________________ and the scientific name is ________________.
13. This type of carbohydrate is not sweet like other carbohydrates. It forms cell walls in plants and gives plants strength and "crunch."
14. This enzyme has its action specifically for table sugar.
15. This enzyme has its action on the sugars found in milk and cheese.
16. People who lack the lactase enzyme are said to suffer from ________________.
17. __________________ are relationships between living things which may or may not be helpful for each other.
18. In this symbiotic relationship one member benefits while the other suffers.
19. In this type of symbiotic relationship one member benefits while the other is not directly affected.
20. The enzyme which breaks down cellulose is known as ________________.
21. ________________ like cows, sheep and deer, have a symbiotic relationship with microorganisms to digest cellulose. Insects, such as _________ also have ____________________ in their digestive tracts to digest cellulose.
22. The chemical formula for glucose is ________________.

E S A T C A L I K O B A N T A P S M Z A U P V I M T S V X X F U R E I T V Y K S
P M R M V O P H S D N E O K I I W K J E L H Z T Q N K Q B K R F U B C K S U U U
D H B C E T D O U F I T K C S U G H L Z D I K Z P C C D S F L D M U X P D C R W
J H G G Z I P C F Q W V N O K W E P B Q E E Z G M Z M C B E C Z I G W J R N Y A
M S I L A S N E M M O C I E I T D Q Y K V N H M R E E U D C X Z N M Y A H S Z S
H Z O U T T Y E I S Z B D I P E L A A T L E F J C W P E E N K T A B S R O M H Q
W L Z Z X C U J M L M U M A L D Q D R F O F D X J M P F W A C R N E B Z U V M Q
H W S U N N B M S Y J K P U X R O F E T F U W J H C A A T R V K T H X W H C Q P
Q C E L L U L O S E Q J N M I L Y C D L P J A Z J A Z R W E Q P S T G A R E J D
H M D U S B X E T A M W J E A U M D N B L K R Y W R A F F L Q D N F L S B A S X
X M R D V E Y W I I V I A K J N N O S G O J O C X B M H T O D N W U R F D C Y C
P R K D H I D I Y N C E C T I C U Z K G L B K I X O Y Y R T G L U C O S E Q C U
A F Z S U K V I S U G A R R E H W J V H L Y S M S H L O A N F R F J N F T T J N
J O O S W X Z G R Z S N Y J O R S Q W Y L O F T J Y O C Q I M S W E M W C O K G
M O N O S A C C H A R I D E D O S E X H E D V M X D S I B E J C E S O T C U R F
R D T L J I L T O S H W Y I V R R Y T V A R B E L R E K J S D G E G D H U X N E
H H G R X I T P C G O C S L A E W G L I P K N N Q A L G D O U E V L M X X W A K
Z T A U Z O M B S I Q A C N W L O E A Z M W L C D T Q W X T W D L Z L I T R F Z
B L H U T A U Z H E C C T A S O W L O N B R G X S E V M H C A D W L S U M K H G
M H I Z Y W U C H C G P A O S T M N A A I A E R E T B J P A O B C N U P L Y Z J
U L K O H A R O H Q Q G I H H E S Z O C L S F T M J X T N L F F L V M B D A W P
Z H L I Q A Y A N V O E C X S M I K L B T B M M Y G Q M L T D E X Y D R U S S U
A Y W V T Q R M E P B A I O D K T J B J R O W S Z T A K D C O M G E O Q O A T E
E E X S G I M H G P D S R G Y A I C T N S A S H N Z T E N R V V F G W P W H E F
Q I E Y D C A D Y I C C P X R J S V I C F G C E E Q T W M O Q N E Z A H N J C Y
G H C E X V H B X G U F J Q P L A A R J J Q W D K U V G P F Z N R W S M H W M K
D L M H D A K Q O S W U P B W F R H K X A D F S F A K T J B H M E T T R M T O B
A I J F B Q I V V V P W J O P K A O Y T J P I N V K R L G N C V H H Q F M Y E L
O K T Y U B C Y N U N L F G O A P N Q S Z R H P Z D Z V O K B E H J C O T L V U
D J A G W D Z W R Y B L S T L F X V H M W U B S U L B V M P G Q A H L R K X M E

Name ____ Date ______________________

Lesson 4 Practice Page 1

Instructions: Fill in the blank with the appropriate word. Refer back to the text portion of the lesson for help.

1. Lipids, like carbohydrates, are formed from three elements: ____________________, ____________________ and ____________________.

2. There are two main portions to a lipid. The first is a three-carbon component known as - ____________________.

3. The second portion consists of chains of carbons and are known as the ____________________- ____________________.

4. There can be up to __________ fatty acid chains per glycerol molecule.

5. A lipid with one fatty acid chain present is known as a ____________________.

6. A lipid with two fatty acid chains present is known as a ____________________.

7. A lipid with ______________ fatty acid chains present is known as a triglyceride.

8. Fatty acid chains can be classified as being __________________ or __________________ depending upon whether any double bonds are present between the carbon atoms making up the chain.

9. Fatty acid chains which have a double bond present are known as __________________ fatty acids.

10. Fatty acid chains which have only single carbon bonds present are known as - __________________ fatty acids.

11. Lipids which are unsaturated usually come from __________________ whereas lipids which are saturated come from __________________ sources.

12. Unsaturated fats are __________________ in form at room temperature while saturated fats tend to be __________________ in form at room temperature.

13. Lipids can be converted into fuel (glucose) for living things through the work of - __________________.

14. Because lipids contain so many carbon, hydrogen and oxygen atoms, the relative amount of energy found in one portion of a lipid is about ____________ times the amount of energy found in an equal portion of a carbohydrate.

15. Lipids are also found to not mix with water. One can then say that lipids are - ____________________ in water or ____________________.

16. Because of their insolubility in water, lipids are useful for living things when water or watery substances need to be contained in one location or another. (True or False)

17. Milk that has all of its fat portion removed is known as ____________________ milk.

18. Milk that comes straight from the cow has a fat content of ____________________.

19. The process whereby the fat portion of the milk (cream) is made to remain equally distributed throughout the watery portion of the milk is known as ____________________.

20. Butter is made from the ____________________ portion of the milk.

21. Explain how butter can be washed with water.

Below are clues to solve the crossword puzzle on the next page.

Across

2. literally meaning afraid of water
3. form of an unsaturated fat at room temperature
4. when one substance mixes readily with another
6. number of bonds carbon atoms desire to form
10. long chain of carbon atoms making up portion of a lipid
12. when one substance does not mix with another substance
13. form of a saturated fat at room temperature
14. source for most saturated fats
15. process whereby lipid portion of milk is equally distributed throughout watery portion of milk
16. three-carbon portion of a lipid
17. source for most unsaturated fats

Down

1. the quantity of energy a fat has versus the quantity of energy found in an equal portion of carbohydrate
2. literally meaning water loving
4. type of fatty acid when only single carbon bonds are present
5. one of three elements found in a lipid
7. type of fatty acid when a double carbon bond in present
8. one of three elements found in a lipid
9. scientific name for a fat
11. one of three elements found in a lipid

Name ____ Date____________________

Lesson 4

Practice Page 2

Instructions: On the previous page, you will find clues to solve this crossword puzzle. Refer back to the text portion of the lesson for help.

1 2 3 4 5 6 7 8 9 10 11 13 14 15 16 17

Name ____ Date______________________

Lesson 5 Practice Page 1

Instructions: Fill in the blank with the appropriate word. Refer back to the text portion of the lesson for help.

1. Like carbohydrates and lipids, proteins consist of the elements: ________________, ________________ and ________________. However, proteins also always contain the element ________________.

2. Proteins consist of smaller components known as __________________________.

3. There are about _______ amino acids known to man of which there are _________ which are required by humans to live.

4. Amino acids consist of two portions: the ________________ portion and then the ________________.

5. The common portion is always the same for each amino acid, but the R-group ____________________ which makes amino acids unique.

6. Of the 20 amino acids necessary for humans to live, many can be made from raw materials in our bodies or other amino acids. However, there are _____ amino acids which cannot be made and therefore must be ___________________________________.

7. List those 8 amino essential amino acids:

8. Amino acids are linked to form ________________. It is the _________________ of these amino acids which form various proteins needed by the body.

9. The process where amino acids are linked together is known as ________________ ________________ because water is released through the process.

10. A ____________________ is formed when two amino acids are linking and a - ____________________ is formed when multiple amino acids are joined together.

11. ______________________ are responsible for dismantling other proteins to provide a source for components to build the amino acids we need. These ________________ also build these new amino acids.

12. The process of breaking down proteins is known as ________________________ which literally means cutting with ________________.

13. Proteins are very large molecules. (True or False.)

14. In general, the primary use for proteins in living things is that they serve as _________________ components for the cells of living things.

15. A common protein found in egg whites is _________________. This protein is also found in our _________________ and is responsible for maintaining appropriate amount of _________________ in our bodies.

16. Another important protein found in our blood is _________________ which is responsible for _________________________ formation.

17. And yet another very important group of proteins are the __________________________________ where the prefix immuno– refers to the _________________ system and globulin refers to _________________.

18. Immunoglobulins work to fight off ___________________________ encountered by the living thing.

19. In mammals, the first milk produced by the mother is full of _________________ and has a special name known as _________________.

Name ____Date____________________

Lesson 5 Practice Page 2

Instructions: Below you will find clues to solve this word find puzzle. Refer back to the text portion of the lesson for help. Note that the words may read forward, backward, up, down or at a diagonal.

1. Proteins are formed from the elements ____________, ___________. ___________ and ___________.
2. Proteins are made of smaller units known as ______________________.
3. Amino acids are made with two main components: the ____________________ and the ______________________ which varies.
4. Amino acids get linked together into peptides through the use of enzymes known as ______________.
5. The bonds which form the linkages between amino acids are known as ______________________.
6. Two linked amino acids are known as a ____________________ whereas many linked amino acids are known as a ______________________ where poly– means many.
7. The joining of amino acids whereby water is released Is known as _________________________.
8. It's the ____________________ of amino acids which determines which protein is produced.
9. The breaking apart of proteins to create a supply of amino acids is known as ________________ because water is used to cut the protein.
10. The primary use for proteins in living creatures is their role as structural components. A protein found in blood is ______________ aids in blood clotting.
11. Another blood protein is ________________ which is the same protein found in egg whites.
12. A final group of proteins are those found in our immune system which are used to fight diseases. This group of proteins are known as the ____________________________.
13. Mammals provide a "megadose" of immunoglobulins through the first milk made for their offspring. This first milk is known as ____________________.

N E N U Y Z E M C P A P X B T V C F J H B D Z P A U O M Y X C T R B R D A G O P

C O O E E S J E E F Y U V C F M A N W K O O E T D F G L M E N J R W E B I G B Q

E J I D G R O P V Y V S X L M M R U L S Q Y I K P A N S C P W P H H Z E R E Z J

P B L T P O T M U R T S O L O C B E O P T F J W B F X X D R T W Y C L L W E J P

J F C I R I R I T G Q E F Z I Q O I J F I W L E C O H A P I N D E N Z Y M E S C

L S T M D O O D G Q Q J N K F Y N K O E Z T B C G E Q G F C R O N P O T P R S L

K Y C A V X P V Y B W F R I X N C W Z M V X D K K P H J A A G Y V O U N I R Y U

O H S N Y D F N F H O U A L R D I S T D B T G B C J L Q T L S G D I Y E J L P U

B E Y G Q E W A O F D W M U E B B K S M A M I N O A C I D S B Y I H X G M N P S

E L E V U N W W I M P U O R G R I N O W D T Z M Z K O P Z O S U X V A O M I M Q

E N R H W X S X J W M W Q P Y M K F N R B Q L J T N M I L I M G M D F R G Z I V

K C S A Z Z F D Y N Q O X U R L X S K B A V X C S I S D E E B J Q I O T X G V O

R C J H U B V Y M B U F C S U S P M R L O M N Y X J N D Z O X I O L N I M C B Z

G A X U B C O P R N P Q B A F Q W I P X C T N C M J E F X Y K C X H K N A P A D

M S Q L C G B D K F H W M N Z J O T Z J S T B A O K H R A B O M Y P I G R O J R

Z E G D O W Z N S V A T S X V J Z X O J H Z L X T Q I X X V T S U A J T C U Y K

F I A O X I B O M L Z N F D L O W X E E S M A D Y U O K U O N Z B A O X M N X N

U N N C W U K B T L V G I Q S K V G S T A S N I L U B O L G O N U M M I B T A H

H I Y P G Q T E I Q W E W Y O O R I X O C S D E Q Z X O S M L C T M F N Q A P R

O Y J K U C G D L Y M B I L E N S O K U Y F F Y K W C S K K B D D F X E U E F V

I A D D U A Y I D U G V W D M T Y I E S H V G Z D E Q Y M C J R H Q Z R T N V V

P J S R M T M T A X O S I H F O L L H F D O E R M S O J E K H E Y M V N H Q X T

E B U C O N H P B W U T D E V W D G I F W Z M X N D A Q M U S T X A M A C L O X

P C O X Y L D E L R P O C P G Q R O Q I P A W Y Y W Y Z B V G A Q Q T W Z L C Y

T S L Q P A Y P F E S N P R J E V N L C O B C P D J F P E C Q W L M D A B P A U

I Y R W N E B S P Q E N O Y I T F K V R J L Q V H Q D G R R U K B L B C P A B C

O B O U I H I Y I U N B J K M U S G R N X P Z O A V T D D I D O B C I P K Y K Y

E W K X D P L K Q S I P E Y E A M X Z T Z K T Y H E O I X I N S F E V N V O G P

Y J Q I H O Z E Y X A N J Z B U P L O Y O N Q B J B N C H E R G H B K D R B O U

U W P H P L S K W J B I Z M L I V S Q I J S B D A V X G B R S Y O R M P F E N E

Name ____Date______________________

Lesson 6 Practice Page 1

Instructions: Fill in the blank with the appropriate word. Refer back to the text portion of the lesson for help.

1. All of the uses for carbohydrates, proteins and lipids in living things, whether the building or breaking down of these molecules, are dependent upon the presence of ________________.

2. These enzymes are very dependent upon the _________ of the environment in which they are working.

3. With regard to the symbol used for pH, the lower case p is thought to represent the term ____________________ while the H stands for ______________________.

4. Substances which have a desire to get rid of hydrogen ions are said to be ___________.

5. Substances which have a desire to accept hydrogen ions are said to be _____________.

6. The pH scale tells us the relative strength of acids and bases. As one moves from value to value on the pH scale the relative strength is multiplied ________ times and the scale is referred to as being a ____________________ scale.

7. An acid with a pH of 3 would be ____________ times stronger than an acid with a pH of 5.

8. On the pH scale, acids are less than _______________ while bases are greater than _____________. Bases taste ________________ while acids taste ________________.

9. Secretions in our stomachs are considered strong acids. What value on the pH scale would you expect if you measured the pH of these gastric secretions? _____________

10. Vinegar or __________________ is a ________________ acid but still is effective at preserving foods. This is because the ____.

11. Write the pH of the following body substances: Blood __________ Milk ____________ Pancreatic secretions __________ Cerobrospinal fluid ________ Urine __________

12. By raising or _________________ the pH of fluids of living things we can directly affect the enzymes functioning in that living thing.

13. pH can be measured by colored strips of paper which have a powder affixed to them. These strips of paper are known as ______________________.

14. Red litmus paper will turn ___________ in the presence of a base.

15. Blue litmus paper will turn ___________ in the presence of an acid.

Name ____Date______________________

Lesson 6 Practice Page 2

Instructions: Below you will find clues to solve this crossword puzzle.

Across

4. base used to leaven cookies

6. greater than seven on the pH scale

8. acids taste

9. bases taste

11. the pH scale's difference between values

13. pH of blood

14. color of litmus paper in presence of acid

Down

1. scientific name for vinegar

2. what the H stands for in pH

3. pH of urine

4. color of litmus paper in presence of base

5. pH strips capable of testing range of pH values

7. less than seven on pH scale

10. enzymes are dependent upon _____ to function correctly

12. what the p stands for in pH

Name ____Date_______________________

Lesson 7 Practice Page 1

Instructions: Fill in the blank with the appropriate word. Refer back to the text portion of the lesson for help.

1. The study of cells is known as ___________________________.

2. The scientist who first made the discovery of cells was ____________________________ in _______________. He gave cells their name because it is said he though they resembled ___________________________.

3. The "little organs" within cells are known as ____________________________.

4. Cells are covered with a "skin" known as the _________________________ or ________________________________.

5. The cell membrane consists of a _______________________________ bilayer.

6. The phosphate groups are ____________________________ while the lipid layer is ___________________________.

7. This allows the membrane to be able to control the passage of ___________________ .

8. Cell membranes are said to ______________________________ meaning that the cell has the ability to allow certain substances to pass but not others.

9. Plants have an extra layer of protection around their cells. This layer is known as the ______________________.

10. Cell walls are made of ________________________ which are long strings of ______________ molecules.

11. Cells walls function like ____________________ for plant cells giving them strength and support.

12. Trees which have thin cell walls and tend to grow quickly are known as ________________________ while trees which have thick cell walls and tend to grow slowly are known as ______________________.

13. The cell organelle which functions like the brain of the cell is the ____________________.

14. The nucleus, depending upon which creature it is in, may or may not have a covering membrane. This membrane is known as the _______________________ or _________________________________.

15. Organisms which DO have a nuclear membrane are identified as being ___________________________ .

16. Organisms which do NOT have a nuclear membrane are identified as being ______________________________.

17. An example of an organism which is a prokaryote is a ____________________________.

18. The structures within the nucleus of a cell which hold the instructions for the cell are the ___________________________________.

19. Organisms are specific for the ___________________ of chromosomes they have and for the information found on each chromosome.

20. Humans have __________ pairs of chromosomes for a total of _________ chromosomes in most of their cells.

21. Within chromosomes we find subunits or "recipes" which are known as ___________________.

22. Genes code for ______________________ in the organism.

23. The organelle of cells which functions like a power plant is the ____________________.

24. The fuel utilized by the power plants of cells is ________________________.

25. The chemical formula for glucose is __________________________.

26. Mitochondria are shaped like ___________________.

27. The inner membrane of a mitochondrion is known as _____________________ and this is where _______________________ takes place.

28. Respiration is the process whereby glucose is converted into _______________ which is in the form of __________s.

29. ATP is the abbreviation for ____.

30. In the presence of oxygen, one glucose molecule can yield ____________ ATPs.

31. This form of respiration is known as _______________________ respiration.

32. Where there is little or no oxygen present, one glucose molecule can only yield _________ ATPs.

33. Respiration with little to no oxygen present is known as ______________________ respiration.

34. The two main uses of ATPs are to allow for _______________________ and _________ production.

Name ____ Date ______________________

Lesson 7 Practice Page 2

Instructions: On the next page, you will find clues to solve this crossword puzzle. Refer back to the text portion of the lesson for help.

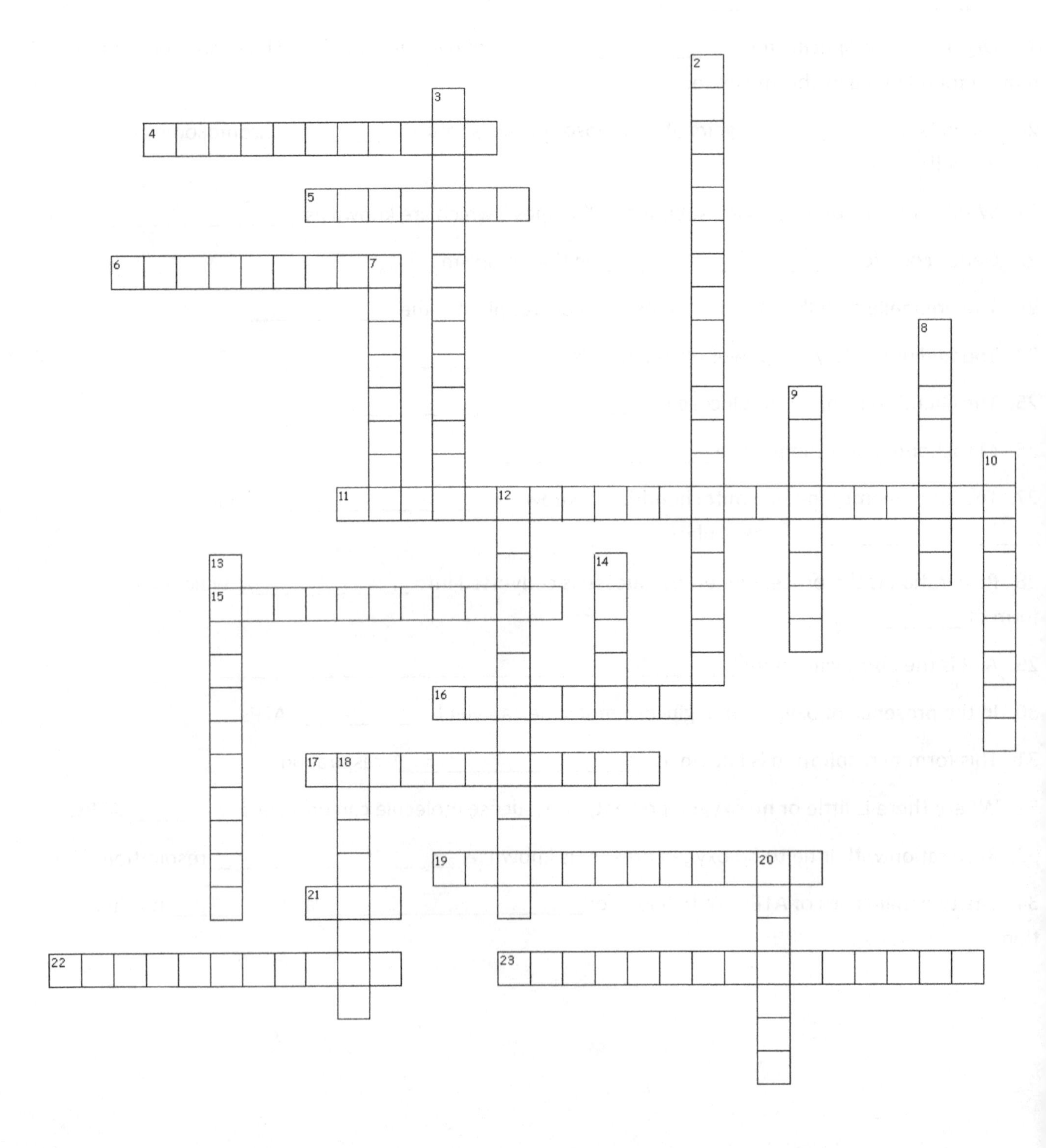

Clues for Crossword Puzzle

Across

4. loves water

5. fuel used by mitochondrion of cell

6. what discover of cells thought they looked like

11. what ATP stands for

15. hates water

16. total number of chromosomes in most of a human's cells

17. aerobic respiration makes this many ATPs.

19. "skin" of cells

21. anaerobic respiration results in this many ATPs.

22. process whereby glucose is converted into ATPs.

23. chemical formula for glucose

Down

2. scientific term which describes how cell membranes are composed of two layers of phosphates and fat compounds

3. "power plant" of cell

7. thin cell wall of pine or fir

8. skeleton of plant cell

9. study of cells

10. carbohydrate which makes cell wall

12. means: allows some things to pass but not others

13. colored bodies which are like chapters in the cell's cookbook

14. "recipes" found in the chromosomes

18. thick cell wall of oak or maple

20. “brain” of cell

Name ____ Date ____________________

Lesson 8 Practice Page 1

Instructions: Fill in the blank with the appropriate word. Refer back to the text portion of the lesson for help.

1. The organelle of cells which is responsible for packaging of products for export from the cell is the ______________________________.

2. When a portion of golgi body is filled with product and ready to be shipped from the cell, a portion known as a ____________________ breaks free to moves toward the periphery of the cell.

3. The organelle of a cell which functions primarily as storage or, in some cases, locomotion, is the ______________________________.

4. The part of a cell which functions like a transportation system is the ________________________ or ________ for short.

5. There are two types of ER: ________________________ and __________________ ER.

6. The roughness is rough ER is due to the __________________________ present.

7. The function of ribosomes are to build ____________ from available ____________________.

8. Another role performed by ER is the folding of ____________________________.

9. There are four degrees of protein folding: ____________________, ____________________, ________________________ and ____________________.

10. The functionality of a protein is dependent upon the way it is ________________.

11. Sarcoplasmic reticulum is found in ______________________ cells and is important in the regulation of ____________________.

12. ______________________________ are like a microscopic skeletal system for cells. They can enable movement, too.

13. Microtubules which make up centrioles are important in ________________________.

14. ______________________________ are organelles found in green plants which are capable of the process of ________________________________.

15. The raw materials for photosynthesis include: ____________________, ____________, and energy from the ________________________.

16. The product of photosynthesis is ________________________ which has the chemical formula of ____________________.

17. This glucose can be used by ______________________________ or can be "stolen" by another living creature to be used as a ______________________.

18. A waste product from cellular ____________________ is ____________________ which becomes a resource for plant use in photosynthesis.

19. Plants and creatures which undergo cellular respiration are __________________ upon each other.

Clues for crossword puzzle on next page.

Across

6. fourth degree of protein folding

9. transportation system of cell

12. second degree of protein folding

13. ER with ribosomes

17. formula for glucose

20. little protein building factories found on rough ER

Down

1. photosynthesis factory of plants

2. reticulum of muscles

3. closet for cell

4. process of making glucose from carbon dioxide, water and sunlight

5. this element is controlled by sarcoplasmic reticulum

7. third degree of protein folding

8. we are ______ upon plants and they are _____ upon creatures like us

10. works like a UPS delivery person

11. tiny bones of cells

14. first degree of protein folding

15. byproduct of photosynthesis but useful for animals

16. important in cell division

18. ER without ribosomes

19. package ready for shipment from cell

Name ______________________________ Date ____________________

Lesson 8

Practice Page 2

Instructions: Find clues to solve this puzzle on the previous page.

Name ____Date______________________

Lesson 9 Practice Page 1

Instructions: Fill in the blank with the appropriate word. Refer back to the text portion of the lesson for help.

1. Living things which are made up of many cells grow in size by ______________________ the number of cells from which they are made.

2. The process of cell division is known as ________________________.

3. The phase of a cell's life cycle when it is going about its assigned job is known as

______________________________.

4. In animals and humans stimulation to divide usually comes in the form of a

________________________.

5. Hormones are substances usually made by __________________ in one part of the body and have their effects in other parts of the __________________ being carried there by the

_______________________.

6. Growth hormone is made by the _________________________ which is located

___________________________________.

7. The cell that receives the signal to divide is referred to as the _________________ cell and the two cells which result from the division are known as the ___________________ cells.

8. The first stage of mitosis is known as _______________________.

9. In prophase the _______________ which was originally uncoiled while being used by the cell, ____________________ or bunches up and becomes visible again as chromosomes.

10. The ____________________________ also begins to break apart during this stage.

11. The second main event of prophase is that the chromosomes of the parent cell must

____.

12. When copying of the chromosomes does not happen correctly, a _________________ is said to have occurred.

13. The second stage of mitosis is known as __________________________. During this phase the chromosomes align themselves along a _______________ within the cytoplasm of the cell. ________________ fibers attach themselves to the chromosomes.

14. The next stage of mitosis is known as ____________________________.

15. During anaphase, the spindle fibers ____________________ pulling their attached

________________ to opposite poles of the cell.

16. The final stage of mitosis is known as ________________________.

17. During telophase the chromosomes begin to ____________________ and the cell membrane begins to _____________________ eventually allowing for the cell to _________ into _____________ cells.

18. The process whereby the cell membrane presses inward to create two daughter cells is known as _________________________.

19. The two "new" daughter cells return back to ____________________ where they continue their jobs.

Name ____ Date______________________

Lesson 9 Practice Page 2

Instructions: Below you will find clues to solve this word find puzzle. Refer back to the text portion of the lesson for help. Note that the words may read forward, backward, up, down or at a diagonal.

1. Phase of mitosis where chromosomes are moving to opposite poles of cell.

2. Part of chromosome where spindle fiber attaches.

3. When chromosome is uncoiled and being used by the cell, this genetic material is called ____________________.

4. During prophase, the uncoiled chromatin will ___________ to form chromosomes.

5. ______________ of the chromosomes are made during prophase.

6. During telophase, the cell membranes pinches inward, a process known as ___________.

7. Parent cells divide into two _____________.

8. Phase of cell life cycle where "regular job" is taking place.

9. Stage of mitosis where chromosomes line up along central plate or equator of cell.

10. Cell division = ______________

11. __________________ divide to form daughter cells.

12. During metaphase, the pairs of chromosomes line up along a central ____________ or equator of the cell.

13. During anaphase the chromosomes get pulled to opposite ________ of the cell.

14. During ____________, chromatin coils up to form chromosomes and the chromosomes duplicate themselves.

15. ____________________ attach to chromosomes during metaphase to eventually pull the chromosomes to opposite poles of cell.

P N F S S B S L P P V A T O E

R E H P P L G I U O N N D H S

O R C I M V L L S A L A Y E A

P G X N C E I E P O U E R E H

H S N D P O T H C G T E S W P

A E P L C J A A H T M I A X R

S I X E L S W T P O N K M O E

E P I F E X E K R H L E E J T

M O A I G R T T N S A H R S N

W C J B C L N F V E X S C A I

H T V E T E E S A H P L E T P

L J L R C C C H R O M A T I N

Y L C S Q S G C G P L A T E H

S H T E X N A V Z G H W P S D

S B S I S E N I K O T Y C R T

Name ____ Date ______________________

Lesson 10 Practice Page 1

Instructions: Fill in the blank with the appropriate word. Refer back to the text portion of the lesson for help.

1. Chromosomes must __________________ themselves in order for each daughter cell to have its own set of chromosomes.

2. Duplication of chromosomes is made possible through the function of _______________ which work to either ________________ or "matchmake" the appropriate parts of the DNA.

3. DNA is the abbreviation for ____.

4. The carbohydrate component of DNA is a ______________________ sugar.

5. The nucleotide portion consists of the ________________________ sugar, a __________________ group and then a pair of ______________________.

6. The "rung" portion of the DNA ladder is made up of pairs of ______________________.

7. There are two groups of bases: the _____________________________ and the _________________________________.

8. The two purines are _______________________ and ________________________.

9. The two pyrimidines are _______________________ and ________________________.

10. _______________________ always bonds with _______________________ while ____________________ always bonds with ______________________.

11. The bonds between the bases are weak bonds known as __________________ bonds.

12. Enzymes _______________ the hydrogen bonds exposing the base pairs.

13. Matchmaking enzymes bring in __________________ in the area joining them according to the matching rules.

14. Eventually, ____________ new _______________ strands of DNA are formed; one strand for each _____________________ cell.

Name ____ Date ________________________

Lesson 10 Practice Page 2

Instructions: On the next page, you will find clues to solve this crossword puzzle. Refer back to the text portion of the lesson for help.

Across

1. always bonds with guanine

3. group of bases containing adenine and guanine

5. always bonds with thymine

6. "new" cells which each need their own copy of the DNA cookbook.

7. Chromosomes must be _________ so that each daughter cell gets a complete copy of all DNA.

8. These are the recipes within each chapter of the genetic cookbook.

10. The sugar found in DNA.

11. The stronger bond between the ribose sugar and phosphate group of a nucleotide is a _______ bond.

12. group of bases containing cytosine and thymine

13. These are the chapters within the genetic cookbook.

14. Part of nucleotide with phosphorus present.

15. these unzip DNA to expose the base pairs

16. Parts of nucleotide.

17. Abbreviation for deoxyribonucleic acid.

Down

2. Shape of DNA discovered by Watson and Crick.

4. Type of bond found between base pairs that is unzipped by enzymes.

9. Chromosomes are made of long strands of _____.

Name ____ Date______________________

Lesson 11 Practice Page 1

Instructions: Fill in the blank with the appropriate word. Refer back to the text portion of the lesson for help.

1. The making of proteins is known as protein ______________________.

2. The "recipe" for making a protein correctly is found in one's ____________.

3. There are four nitrogen bases found in DNA and they are ___________________, ____________________, ______________________ and ___________________.

4. And, it is the ________________________ of these bases which determines the type of protein that can be synthesized.

5. Before a protein can be made, the particular section of DNA must be ________________ by ______________________ to expose the base pairs.

6. After the base pairs are exposed, ______________ can be formed to make a "copy" of the sequence of the bases.

7. Unlike DNA, RNA is __________________ stranded and contains the base _______________ instead of __________________.

8. The RNA which forms on the exposed DNA bases takes the name _____________________ or mRNA for short.

9. The formation of mRNA takes place in the ________________________ of the cell whereas protein synthesis takes place out on/in the ________________________________ of the cell.

10. More specifically, proteins get built in the _____________________________ found on the _________________ endoplasmic reticulum

11. Out at the ribosome, available amino acids are "tagged" with _____________________ or tRNA for short.

12. The tRNA is specific for each _____________________ and will therefore only fit the appropriate sequence of the mRNA that has arrived at the ribosome. The tRNA works like a _________________ that fits the mRNA "lock."

13. The tRNA is in sets of _________________ base pairs and is called an _______________.

14. Specific codons code for specific _________________ acids.

15. As the amino acids get aligned to the mRNA, _____________________ work to link together forming ______________________ bonds. Long strands of peptides form in this way.

16. The process whereby mRNA is made from a section of DNA is called _______________________________ where the process where tRNA assembles amino acids in the ribosome is known as _______________________________.

17. When one's DNA is altered or damaged, the proteins made from this damaged DNA may not _____________________________ correctly. A change in the DNA is called a ___________________________ mutation.

18. Things that cause mutations are called ______________________ or __________________.

19. An example of a mutagen is _______________________________ which is energy that can come from the sun or an elemental source.

20. Cells which have had their DNA altered may become _____________________ cells and change their behavior.

21. Cancer treatments may include bombarding cancer cells with radiation in an attempt to cause ____________________ in the DNA of those cells leading to their destruction.

Name ______________________________ Date ________________

Lesson 11 Practice Page 2

Instructions: Below are 17 words or phrases that we've introduced in this lesson. Their spellings have been scrambled. Rearrange the letters to spell these words and place them into the boxes. Transfer the letters in the numbered boxes to the "finale" at the bottom. If you need hints on unscrambling the words, turn the page.

QESNECUE [ ][ ][ ][ ][ ][ ][ ][ ] 16

MGNERSERNASE [ ][ ][ ][ ][ ][ ][ ][ ][ ][ ][ ][ ] 18 24 25

SENFARANTRR [ ][ ][ ][ ][ ][ ][ ][ ][ ][ ][ ] 5 21

NAD [ ][ ][ ] 19

BOIRESMOS [ ][ ][ ][ ][ ][ ][ ][ ][ ] 23 3 20

LUUSECN [ ][ ][ ][ ][ ][ ][ ] 8

MIENIRASOLLUCTEDUCPM [ ] 1 17

GEURHOR [ ][ ][ ][ ][ ][ ][ ] 2 12

ZEYNMES [ ][ ][ ][ ][ ][ ][ ] 22 9

DOPSBDETPENI [ ][ ][ ][ ][ ][ ][ ][ ][ ][ ][ ][ ] 4

MANODSICIA [ ][ ][ ][ ][ ][ ][ ][ ][ ][ ] 6

SONCDO [ ][ ][ ][ ][ ][ ] 10

RANSOITLNAT [ ][ ][ ][ ][ ][ ][ ][ ][ ][ ][ ] 7

SRNAITOINTRPC [ ][ ][ ][ ][ ][ ][ ][ ][ ][ ][ ][ ][ ] 11

TOTMANSIU [ ][ ][ ][ ][ ][ ][ ][ ][ ] 14

ANRIATIOD [ ][ ][ ][ ][ ][ ][ ][ ][ ] 15

REACNC [ ][ ][ ][ ][ ][ ] 13

[1][2][3][4][5][6][7] [8][9][10][11][12][13][14][15][16] [17][18] [19][20][21][22][23][24][25].

Hints for solving the scrambled words puzzle are found below. Note that these hints are not in the same order as the puzzle words.

Transcribing molecules

These get linked together to form proteins.

It is the ________________ of bases which determines the protein synthesized.

These make smooth ER not smooth.

Sets of threes.

Location of DNA

ER

ER that is not smooth.

Tags or keys found on amino acids

Scissors or matchmakers

Bonds between amino acids.

When genes get messed up.

Rays from the sun.

Disease where cells are out of control.

Ladder-like molecule with all information.

mRNA does this job

tRNA does this job

Name ____Date______________________

Lesson 12 Practice Page 1

Instructions: Fill in the blank with the appropriate word. Refer back to the text portion of the lesson for help.

1. The process by which living organisms create more complete living organisms is known as ______________________________.

2. Organisms with two parents have pairs of _______________________, one member of each pair coming from each ____________________.

3. Humans have _________ pairs of chromosomes for a total of ___________.

4. The variation in a trait is known as an ___________________. Examples of alleles for eye color could be: ____.

5. When it comes to alleles, chocolate is to ice cream as ______________ is to the gene for hair color.

6. Living creatures which have pairs of chromosomes are said to be _______________ or 2N.

7. Almost all cells in humans and animals are 2N except for the _________ cells or ____________________.

8. These cells (sex cells or gametes) are ____N or haploid.

9. The cells which eventually become gametes are known as the __________________ _________________________ and originally are _____N.

10. The primordial sex cells undergo the process of __________________ to reduce the number of chromosomes from 2N to _______.

11. In human males, one primordial sex cell will result in ___________ sperm cells.

12. In human females, one primordial sex cell will result in ____________ ovum and three cells which do ______ survive.

13. Human males continue to create more primordial sex cells throughout their lives, while human females are born with the total number of primordial sex cells they will have their entire lives. (True or False)

14. In males, the process of creating sperm cells takes place in the _______________.

15. In females the process of creating ova takes place in the ___________________.

16. The joining of the 1N sperm cell with the 1N ovum to create a 2N individual is known as _________________________________.

17. Reproduction which involves two parents, a male and female each contributing a single cell for the new offspring, is known as __________________ reproduction.

18. Reproduction which involves only one parent organism is known as ________________ reproduction.

19. In ______________________ reproduction there is a great possibility for an assortment or mixing of genetic information from each parent.

20. In ______________________ reproduction, the offspring carries the exact same genetic information as the parent. There is not any change.

21. Bacteria utilize a means of asexual reproduction whereby the "parent" bacteria creates a copy of its genetic material and then promptly splits into two equal halves, each with its own copy of genes. This method of reproduction is known as ______________________.

22. Another means of asexual reproduction is where genetic material is isolated near the outside of the organism. The cell membrane encircles the genetic material and other cellular organelles forming a bud. The bud eventually breaks free. This method is known as

________________________________.

23. Some plants, like strawberries and spider or airplane plants create "baby" plants by sending out special stems called ______________________ which eventually touch ground forming roots and new leaves. This method that plants use is called __________________________________

______________________________.

24. The plants which result from vegetative propagation will have genetic information unlike the parent plant. (True or False)

25. Ferns and mushrooms reproduce by producing spores. This method of reproduction is known as

_________________________________.

26. Some creatures, like starfish for example, are able to regenerate whole new living organisms from broken parts. This form of reproduction is known as ____________________.

27. Two parent involvement = __________________________ reproduction.

28. One parent involvement = __________________________ reproduction.

29. Sexual reproduction = mixing of genetic information in offspring. (True or False)

30. Asexual reproduction = no mixing of genetic information in offspring. (True or False)

Name ____Date______________________

Lesson 12 Practice Page 2

Instructions: On the next page, you will find clues to solve this crossword puzzle. Refer back to the text portion of the lesson for help.

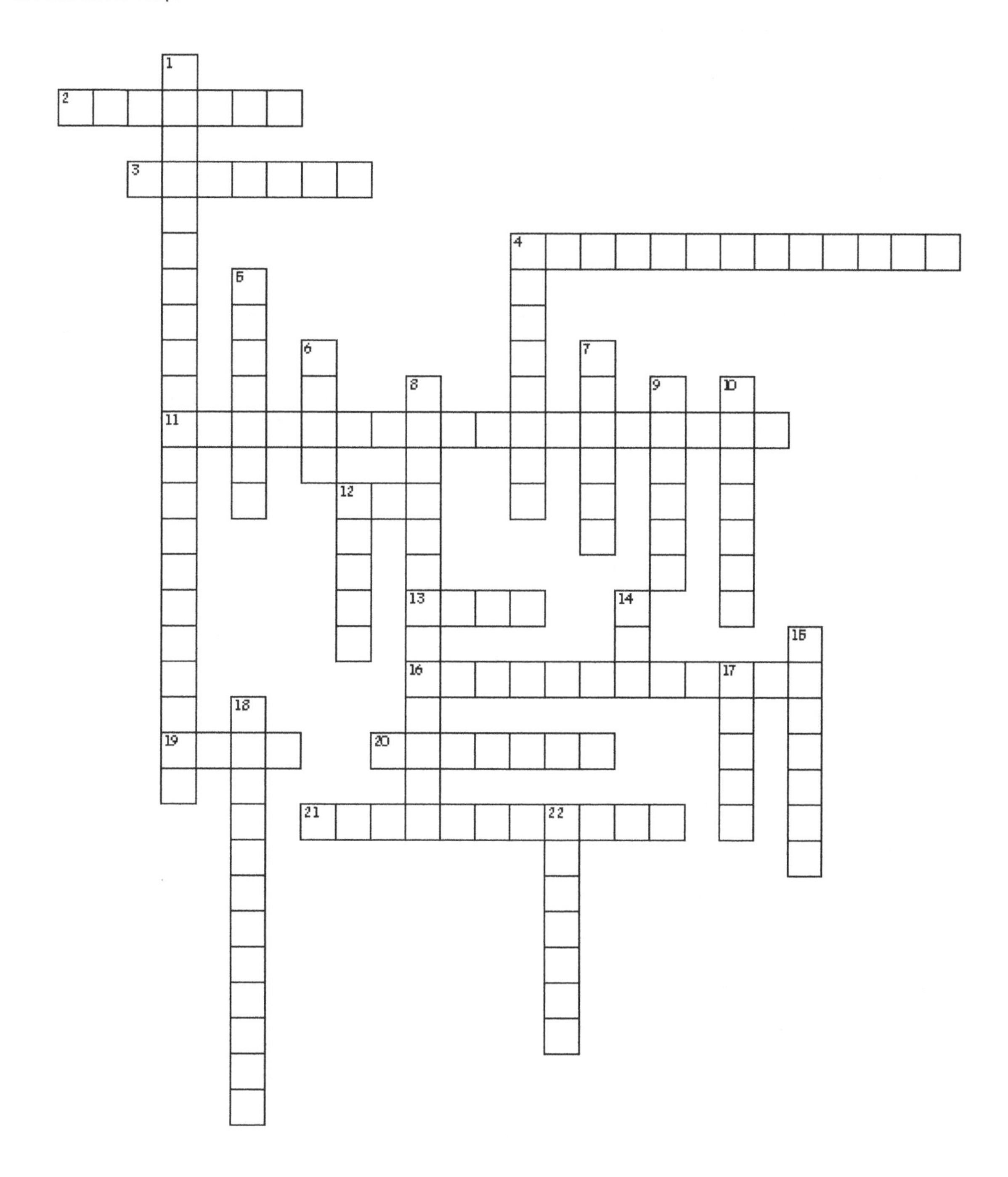

Across

2. term used for sex cells

3. process where a diploid cell becomes haploid

4. joining of ovum and sperm to create new individual

11. cells which are 2N and destined to undergo meiosis to become 1N 12. female gametes

13. number of gametes created in males for each primordial sex cell

16. process of reproducing through the formation of spores

19. haploid

20. having pairs of chromosomes

21. number of pairs of chromosomes in humans

Down

1. means of reproducing often employed by plants where specialized stems or roots grow away from parent plant with small "baby" plant forming on end of stem

4. total number of chromosomes in humans

5. asexual form of reproduction where a small portion of parent cell emerges and then breaks free from parent cell

6. diploid

7. reproduction where two parents are involved

8. form of reproduction where parent cell divides into two new cells

9. location in males where gametes are produced

10. variations of a trait on a chromosome

12. location in females where gametes are produced

14. number of gametes created in females for each primordial sex cell

15. reproduction where only one parent creates new living organism

17. male gametes

18. process where a living organism creates more living organisms like itself

22. having half the number of chromosomes

Name ____Date______________________

Lesson 13 Practice Page 1

Instructions: Fill in the blank with the appropriate word. Refer back to the text portion of the lesson for help.

1. The biologist who discovered many concepts of genetics was named ________________.

2. One of Mendel's biggest contributions was that traits are ________________________

____.

3. The record of genes found in a living organism is referred to as the _________________.

4. The results of the genotype that can be observed whether physical attributes or behaviors is referred to as the _______________________________.

5. The possible variations of a gene are referred to as ___________________________.

6. We can think of alleles as the possible ___________________ available for ice cream or cookies.

7. The Law of _______________________________ says that, in meiosis, one chromosome from each parent primordial sex cell will go to each resulting gamete.

8. The Law of _________________________________ says that traits get distributed independently from other.

9. When looking at the genotype of the offspring that was created through sexual reproduction, we find ____________________ present, one from the male parent and one from the female parent.

10. While both alleles are present, only one gets expressed. The ___________________ allele, if present, is one that always gets expressed.

11. The allele that is present, but does not get expressed is the __________________ allele.

12. When describing genes, one usually uses ________________________. To indicate dominant alleles, one uses ____________ case letters and recessive alleles are indicated by ________________ case letters.

13. When dominant alleles are inherited from both parents, for example HH, the individual is said to be ____________________ dominant for the H allele.

14. When recessive alleles are inherited from both parents, for example hh, the individual is said to be ____________________ ____________________ for the H allele.

15. When an individual inherits a combination of a dominant and recessive allele for a specific trait,

that individual is said to be ______________________________ for that trait.

16. Suppose hair color is indicated using the letter "H." Suppose, also, that black hair color is dominant and red hair color is recessive. If this is the case, what color hair would these individuals have: HH ________________________ Hh__________________ hh__________

17. If Tom has black hair (from question 16 above,) what are his possible genotypes?

__________________ or ______________________

18. If Tom has red hair, what is his only possible genotype? ____________________

19. If Tom is heterozygous for black hair color, what must his genotype be? ______________

20. If Tom is homozygous dominant for black hair color, what must his genotype be? _____

21. If all of Tom's biological children have black hair, regardless of the hair color of his wife, what must Tom's genotype be? ____________________

22. If Tom has some children with black hair and some with red hair, what must his genotype be?

____

23. If Tom has the genotype Hh and his wife has the genotype Hh, do they have the chance of having a child with red hair? __________________

24. If Tom has red hair and he marries his wife who has black hair, could they ever have children with red hair? _______________ How can this be? ________________________

25. If Tom has red hair and he marries his wife who also has red hair, will they ever have children with black hair (according to our "story" of hair color dominance)? _________

26. If Tom has black hair and his wife has black hair, could they ever have the chance of having a child with red hair? ________________ If so, how could this occur? __________

____.

27. Suppose Tom is homozygous dominant for black hair color and his wife is homozygous dominant for black hair color. Will they ever have the chance of having children with red hair?

_______________ Why or why not? ____.

28. The situation where the phenotype of an individual appears to be the result of sharing of two dominant alleles is known as ____________________________.

29. When one finds that certain traits, such as down color of chicks, is found only in one gender, the trait is said to be ___________________________.

Name ____ Date ______________________

Lesson 13 Practice Page 2

Suppose you have some show-quality cats that you would like to breed to be able to sell their kittens. While all of the kittens bring good profits, the ones with long hair bring the most returns. Suppose that hair length in cats utilizes the letters "L" and "l" for the genotype. Suppose, also, that long hair length is a recessive trait. Look at the following situations and predict the possible outcomes for hair length in your kittens. Use the Punnet squares to support your predictions. Male cats are referred to as toms and females are referred to as queens.

A. Your tom is homozygous recessive (l l) and your queen is heterozygous (Ll) for hair length.

Number of kittens with

Long hair: ___________

Short hair: ___________ out of every four kittens.

B. Your tom is heterozygous (Ll) and your queen is homozygous (l l) recessive for hair length.

Number of kittens with

Long hair: ___________

Short hair: ___________ out of every four kittens.

C. Your tom is heterozygous and your queen is heterozygous for hair length.

Number of kittens with

Long hair: ___________

Short hair: ___________ out of every four kittens.

D. Your tom is homozygous recessive and your queen is homozygous recessive.

Number of kittens with

Long hair: ___________

Short hair: ___________ out of every four kittens.

E. Your tom is homozygous dominant and your queen is homozygous dominant.

Number of kittens with

Long hair: ___________

Short hair: ___________ out of every four kittens.

Which situation (question letter, A-E) would potentially be the best money-making opportunity for you? ______________________________

Name ____Date______________________

Lesson 14 Practice Page 1

Instructions: Fill in the blank with the appropriate word. Refer back to the text portion of the lesson for help.

1. The study of the form of living things is known as ____________________________.

2. The study of grouping organisms based upon similarities and differences is known as

_______________________.

3. While many scientists have contributed to the study of taxonomy, one in particular can be credited for a tremendous amount of early work. His name was ________________.

4. The most important reason living things are placed into groups and given specific names, is that scientists need to ___________________________ accurately about what they learn about living things with each other.

5. Living things all fall into one of five major groups known as kingdoms. These five kingdoms are:

_______________________, _______________________, __________________,

_________________________ and _______________________.

6. Examples of the Animalia kingdom might include: ________________, _____________ and

_______________________.

7. Examples of the Plantae kingdom might include: ________________, _____________ and

_______________________.

8. Examples of the Fungi kingdom might include: ________________, _____________ and

_______________________.

9. Bacteria and viruses are members of the _______________________ kingdom.

10. Unicellular, swimming creatures are members of the ____________________ kingdom.

11. The classification division immediately beneath the kingdom level is the _____________ level.

12. Humans fall into the phylum ______________________ because of the presence of a

_________________________.

13. Why do dogs also belong to this phylum? _____________________________________

14. Beginning with kingdom and phylum, list the remaining levels of classifications for living things: kingdom, phyum, __________________, __________________, ____________,

_____________________ and ___________________.

15. The level of classification which serves as the "first name" of an organism is __________ while the

____________________ level serves as the "last name."

16. There are rules for writing the scientific name of organisms. The ________________ portion is capitalized while the ________________ portion is not. If hand-written, they are to be ______________________________ and if typed, ________________________.

17. In this lesson, we found that size was a determining factor used in grouping in members of the cat family. In horses and cattle, a determining feature was the number of ______ on each foot. Cattle are classified as being ____________________ while horses are classified as being ____________________. 18. How would a whitetail deer be classified? ______________________________. How about a gray zebra? ______________________

19. Using the Catalogue of Life resource, find the scientific names for the following living organisms:

A. Eastern Cottontail rabbit: ____________________________________

B. White oak: ____________________________________

C. Oyster mushroom: ______________________________

D. Domestic dog: ____________________________________

E. House fly: ____

F. Red squirrel: ____

20. Using the Catalogue of Life resource, find the common name for the following living organisms:

A. *Bos taurus* ________________________

B. *Sus scrofa* ________________________

C. *Canis lupus* ________________________

D. *Julglan nigra* ________________________

E. *Homo sapien* ________________________

Name ____Date______________________

Lesson 14 Practice Page 2

Instructions: Below, you will find clues to solve this crossword puzzle. Refer back to the text portion of the lesson for help.

Across

3. Kingdom that we find puffballs.

4. Fellow who did much of early taxonomy.

6. Kingdom of tiny swimmers.

8. Chordata is an example

11. All members of the phylum Chordata have _______________

13. Lions and tigers and bears, oh my!

15. study of groupings of living things based upon similarities and differences

Down

1. "highest" category of taxonomic hierarchy

2. Sunflowers belong to this kingdom.

5. Species name is written using _______________

7. Scientific name format when type-written.

9. Genus name is always written using _________

10. Kingdom that we find bacteria.

12. Phyla are divided into _____________.

14. study of forms and shapes of living things

Name ____ Date ________________________

Lesson 15 Practice Page 1

Match the word or phrase in Column A with the <u>best</u> word or phrase in Column B. Write the letter from Column B beside the number in Column A. The first question has been completed for you!

__A__1. Kingdom Animalia

_____2. Rotifera

_____3. Phylum Mollusca

_____4. Class Cestoda

_____5. Characterized by having namatocysts

_____6. Sea anemones belong in this class

_____7. Flukes are a member of this class.

_____8. Flat worms

_____9. Comb jellyfish belong to this phylum.

_____10. True jellyfish

_____11. Phylum Annelida

_____12. Phylum Bryozoa

_____13. Phylum Brachiopoda

_____14. Sponges

_____15. Class Bivalvia

_____16. Class Gastropoda

_____17. Class Diplopoda

_____18. Chitons

_____19. Phylum Acanthocephala

_____20. Class Insecta

_____21. Class Arachnida.

_____22. Subphylum Crustacea

_____23. Class Chilopoda

_____24. Class Cephalopoda

_____25. Phylum Arthropoda .

_____26. Phylum Echinodermata.

_____27. Phylum Hemichordata.

_____28. Phylum Chordata

A Includes all animals

B Phylum Platyhelminthes

C Phylum Porifera

D Phylum Cnidaria

E Class Scyphozoa

F Class Anthozoa

G Class Trematoda

H Have a little crown of twirling cilia.

I Phylum Ctenophora

J Tapeworms

K Spiny-headed worms

L Moss animals

M Upper and lower shell

N Have a mantle which creates a shell

O Includes snails, slugs and whelks.

P Clams, oysters and scallops

Q Members include the squid and octopus

R Shell made of eight valves known as plates

S Segmented worms like earthworms

T Segmented body parts and jointed legs.

U Spiders

V Two pairs of antennae and ten legs.

W Centipedes

X Insects

Y Millipedes

Z Pentaradially symmetrical.

AA Acorn worms

BB Animals with backbones.

Name ____ Date ______________________

Lesson 16 Practice Page 1

Match the word or phrase in Column A with the best word or phrase in Column B. Write the letter from Column B beside the number in Column A. The first question has been completed for you!

Column A	Column B
__A__1. Phylum chordate	A Phylum of animals possessing a notochord, gill slits and a postanal tail (at one point in development).
_____2. Subphylum Cephalochordata	B Lancelets
_____3. Class Chondrichthyes	C No-jawed fish; lamprey and hagfish
_____4. Order Trachystoma	D Sharks, rays and skates
_____5. Class Osteichthyes	E Fishes that have bony skeletons
_____6. Order Anura	F Aquatic at some point in their lives and then become terrestrial
_____7. Class Amphibia	G Amphibians with no legs, blind worms
_____8. Order Apoda	H Newts and salamanders
_____9. Class Agnatha	I Amphibians with no tail, frogs, toads.
_____10. Class Aves	J Rough-mouthed amphibians
_____11. Class Reptilia	K Covered with scales, only have lungs, internal fertilization
_____12. Order Rhynchocephalia	L Tuatara of New Zealand
_____13. Order Edentata	M Order include turtles and tortoises.
_____14. Order Squamata	N Includes crocodiles and alligators
_____15. Order Chiroptera	O Snakes and lizards
_____16. Order Crocodilia	P Class includes all birds
_____17. Order Urodela	Q Warm-blooded, milk-producing, hairy animals
_____18. Order Insectivora	R Included the egg-laying mammals: platypus and echidna
_____19. Class Mammalia	S Pouch-possessing mammals
_____20. Order Chelonia	T Moles and shrews
_____21. Order Rodentia	U flying mammals, bats
_____22. Order Monotremata	

_____25. Order Cetacea

_____26. Order Carnivora

_____27. Order Sirenia.

_____28. Order Artiodactyla

_____29. Order Proboscidae.

_____30. Order Lagomorpha

_____31. Order Pinnipedia

_____32. Order Perissodactyla

_____33. Order Primates

V Peg-like teeth armadillo and sloth

W Includes whales, porpoises and dolphins.

X Members of this order include the manatees

Y Elephants

Z Meat-eaters; foxes, wolves, dogs, cats

AA Seals, walruses

BB Horses, donkeys, rhinos

CC Include the cattle, sheep, deer, giraffe

DD Monkeys, lemurs, gibbons, orangutans, chimpanz gorillas and humans.

EE Includes rabbits, hares and the pikas

FF Rats, mice, squirrels

GG The pangolins

Name ____Date______________________

Lesson 17 Practice Page 1

Please fill in the blank(s) for each statement below. Use your lesson to help you.

1. Plants, through the process of ____________________ are able to take carbon dioxide, water and sunlight to produce ______________________ which is fuel for the plant.

2. Plants are considered to be _____________________ in that they are able to produce their own food supply. Living things which depend upon other living things as food are said to be _____________________.

3. Photosynthesis in plants takes place in the _________________ of the cells of plants.

4. Unlike the cells of animals, plants have _______________ which are made up primarily of _____________ which are long chains of glucose molecules.

5. Taxonomists divide plants into two large groups based upon the presence of ___________ tissue.

6. Vascular tissue which carries water is known as the _______________ while vascular tissue which carries food (glucose) around the plant is known as __________________.

7. Instead of being divided into phyla, plants within the Kingdom Plantae are divided into _______________.

8. There is one division of plants in the non-vascular division. This is the Division ___________________.

9. Three classes are present within this division. The plants found in these classes include the ______________________, ___________________ and ___________________.

10. The Division Psilophyta has ___________-like leaves. The most common member is the _____________.

11. The Division Sphenophyta includes plants known as ______________________ which have hollow stems and contain ________________.

12. The Division Pterophyta includes ______________ which have well developed underground rhizomes.

13. Cycads belong to the Division _____________________ and have seeds not enclosed in _____________.

14. The lone member of the Division Gingkophyta is the ______________________.

15. The ____________________________ plant is a member of the Division Gneophyta has two very long leaves and a very short trunk.

16. Trees such as pines, firs and spruces belong to the Division ____________________. They are commonly referred to as ____________________ and have their seeds in ____________________.

17. Their leaves are like ________________.

18. Plants which have flowers and seeds enclosed in a fruit belong to the Division ____________________.

19. There are two classes found within the Division Anthophyta. The first includes plants with one cotyledon and are commonly referred to as ____________________. Examples of these include: ____________________

20. Flowering plants with seeds having two cotyledons are commonly referred to as ____________________ and examples are: ____________________.

21. Flowers with both male and female structures present are known as ______________ flowers whereas flowers with only one sex of reproductive structure are known as ______________ flowers.

22. The male parts of a flower are the stem-like ________________ with the pollen-producing part known as the ____________________.

23. The female part of a flower includes the ________________ which has a stem-like portion known as the ________________ and the sticky top known as the ____________________.

24. Pollen carry two types of cells: one which forms the ________________ while the second divides to create two ____________________.

25. The pollen tube allows the ______________ cells to join the ______________ present in the ovary of the flower. From this joining process, many cell divisions result and the seed, enclosing the plant embryo, develops.

Name ____ Date______________________

Lesson 17 Practice Page 2

Use the clues on the next page to solve this crossword puzzle regarding the plant kingdom.

Clues:

Across

8. plants having parallel leaf veins and seeds with one cotyledon

10. structures within ovary which are female contribution to new plant embryos

11. tissue of plants which is capable of transporting water and food products

13. division of plants which includes flowering plants

14. division of plants which includes pines, spruces and firs

15. living things which depend upon other living things as a food supply

Down

1. organelles which carry out photosynthesis

2. plants with branched leaf veins and seeds with two cotyledons

3. sticky end of female pistil where pollen lands

4. leaf-like covering of a flower

5. process whereby plants create glucose from sunlight, water and carbon dioxide

6. colorful part of flower which attracts pollinators

7. particle which contains cells which create pollen tube and sperm cells

9. plants are considered to be this because they can make their own food.

12. portion of male flower part that produces pollen grains

Name ____Date______________________

Lesson 18 Practice Page 1

Please fill in the blank(s) for each statement below. Use your lesson to help you.

1. Members of the Kingdom Monera are considered to be ________________________ meaning they have no nuclear membrane nor organized organelles.

2. Members of the Kingdom Monera also are ________________________ meaning they solely consist of being one cell.

3. The primary forms of reproduction in the Kingdom Monera is through __________________ and ________________________.

4. The phylum of Monera which includes the bacteria which causes disease is the Phylum ____________________________.

5. Members of this phylum (question 4) live off of ______________________ or live as parasites of living things.

6. Viruses consist of a strand of _______________ or _____________ enclosed by a _________________________ known as the _______________________.

7. Viruses are considered to be ______________________ ______________________ as they must live within the cells of a host to survive.

8. Control of viruses is mainly through the use of _________________ which stimulate the body's immune system to produce ____________________ to fight off the disease when encountered at a later time.

9. Viruses work by entering a host cells and taking over __________________ of the cell's DNA in an effort to get the cell to make more viruses. The host cell then dies.

10. Viruses are classified as either being DNA or ____________ viruses.

Name ____Date________________________

Lesson 18 Practice Page 2

Use the clues below to solve this crossword puzzle regarding the Kingdom Moneran.

Clues:

Across

3. "living" things which cause disease

4. agent given to stimulate immune system to fight off virus should it ever arrive

6. have no cell membrane nor organized organelles

7. meaning only one-cell-big

Down

1. means of asexual reproduction utilized by monerans

2. kingdom of unicellular, prokaryotes

5. protein covering of a virus

Name ____Date_____________________

Lesson 19 Practice Page 1

Please fill in the blank(s) for each statement below. Use your lesson to help you.

1. Members of the Kingdom Protista are all ___________________________ creatures meaning they are made up of organisms of only one cell.

2. They differ from members of the Kingdom Monera in that they have ____________________ organelles and are therefore identified as being ___________________________ as opposed to the Monerans which are prokaryotic.

3. While there are several phyla of the Kingdom Protista, these can be grouped into three main groups. These are the _______________-like protists, the _______________-like protists and the _______________-like protists.

4. Members of the animal-like protists are grouped based upon _____________________.

5. ____________________ are the false-foot like appendages of Phylum Sarcodina.

6. ______________________ are the hair-like projections found in the Phylum _____________.

7. Members of the Phylum _______________________ move about through the use of a flagellum which is a tail-like structure capable of propelling the organism through the water.

8. Members of the Phylum ______________________ are animal-like yet don't move about. Members include the ______________________.

9. The fungal-like protists include slime molds which move about on the ___________________ eat-eating up dead leaves.

10. The devastating potato famine of the 1800's was caused by a member of the _________________ phylum.

11. The plant-like protists include various phyla of __________________. They are categorized based upon the presence of various colors of ___________________________ as well as how they store _______________________.

12. Chlorophyta are the _____________ algae. Rhodophyta are the _____________ algae.

13. Phaophyta are the ___________________ algae. Chrysophyta are the _______________ algae.

14. Pyrrophyta have the ability to produce light through a process known as _________________.

Name ____Date______________________

Lesson 19 Practice Page 2

Use the clues on the next page to solve this crossword puzzle regarding the Kingdom Protista.

Clues:

Across

2. algae capable of producing light

5. animal-like protists

7. fungal-like protists of forest floor

10. green algae

11. brown algae

14. hair-like projection allowing for movement

Down

1. living thing capable of producing light

3. red algae

4. false foot

6. golden brown algae

8. having membrane-bound organelles

9. tail-like structure of protozoans

12. plant-like protists

13. *Phytophthora spp*. caused this in Ireland in the mid-1800s

Name ____Date______________________

Lesson 20 Practice Page 1

Please fill in the blank(s) for each statement below. Use your lesson to help you.

1. Fungi are ______________________meaning they live off of dead organisms. They are known as ______________________.

2. Can be categorized according to the presence or lack of ____________________ in their hyphae. Hyphae can be considered to be like __________ which grow through the substance being "eaten" by the fungus.

3. Fungi can reproduce ______________ and ____________________.

4. Fungi which belong to the Division Zygomycota have no _________________ in their hyphae. The common ________________________ is a member of this division.

5. Fungi which belong to the Division Basidiomycota have _________________ in their hyphae. Familiar members include ____.

6. Fungi which belong to the Division Ascomycota have cross walls with ___________________and reproduce through the use of ___________or sacs which produce ________________. Morels and _______________ are members of this division.

7. Fungi which belong to the Division Deuteromycota have ______________________________ cycles which are not yet fully understood. Athlete's foot fungus and the fungus used to create ________________ and flavor _________________ belong in this division.

Name ______________________________ Date ________________

Lesson 20 Practice Page 2

Use the clues below to solve this crossword puzzle regarding the Kingdom Fungi

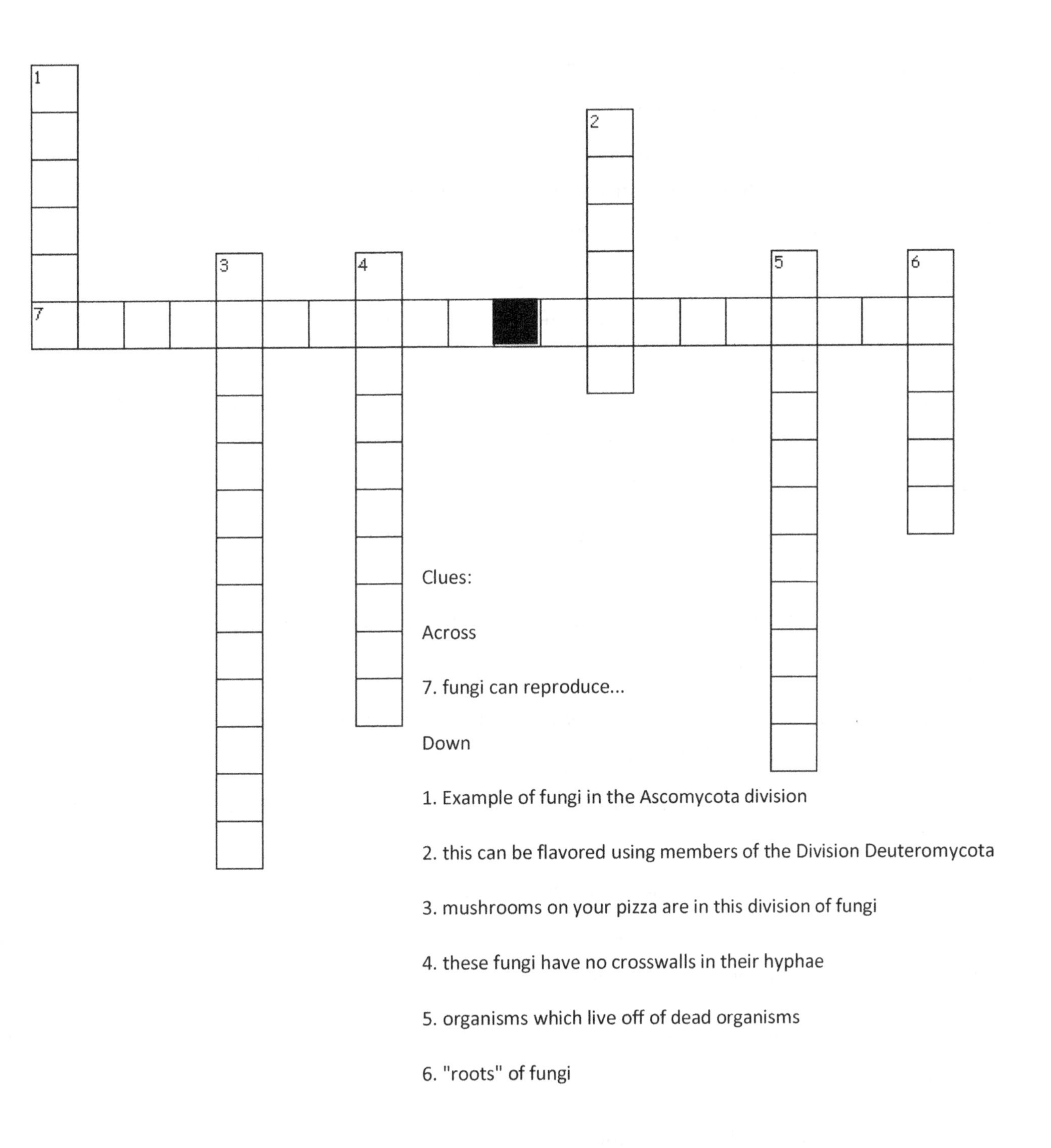

Clues:

Across

7. fungi can reproduce...

Down

1. Example of fungi in the Ascomycota division
2. this can be flavored using members of the Division Deuteromycota
3. mushrooms on your pizza are in this division of fungi
4. these fungi have no crosswalls in their hyphae
5. organisms which live off of dead organisms
6. "roots" of fungi

Name ____Date______________________

Lesson 21 Practice Page 1

Please fill in the blank(s) for each statement below. Use your lesson to help you.

1. The study of how living things are made is known as _________________ while how living things function, is known as _______________________.

2. At birth, a baby has over ________ bones of which several fuse together to reduce the number down to about ____________ in an adult.

3. There are the ___________ bones which consist of the bones making up the arms and legs, the ________ bones which are bones primarily of the skull and pelvis and the ___________________ bones which are the bones of the spinal column

4. Initially, the long bones were formed of ______________ which over time is replaced by _________ and _____________ containing compounds which make the bone hard. This hardening process is called _________________.

5. On long bones, there are two _________________, one found on each end and then one _________________ which joins the two epiphyses.

6. Locations where ossification begins are known as __________________________________.

7. The boundary where growth of the bone occurs is referred to as the _________________or _________________ of the bone.

8. The joints between of the long bones of the arms and legs are known as _________________ joints.

9. If you look at the ends of long bones, you'll find a layer of _________________cartilage.

10. To reduce friction between the ends of two adjoining bones, a fluid known as _________________ keeps the bones gently sliding over each other

11. The tough membranes which bridge the joint space between the two bones is known as the _________________

12. Over time and with age, this slippery joint surface may become worn and result in pain and inflammation. This is known as _________________.

13. The joint between the femur and the pelvis at the hip is classified as a _________________ joint.

14. The joint at the knee and elbow are classified as _________________ joints as the mainly allow for bending in one direction.

15. The cells which make up bone are known as _________________.

16. There are two kinds of osteocytes. The __________________ are bone cells which have the responsibility of building new bone, like along the growth plate, while the __________________ have the job of taking apart or tearing-down bone.

17. Bones are also very important when it comes to __________________ of vital organs in the body. The flat bones of the skull protect the __________________ while the irregularly-shaped bones of the spinal column protect the __________________. The __________________ also are very important for protecting the lungs, heart and upper abdominal organs.

18. The production of red blood cells takes place in the __________________ of bone This tissue is known a hemopoietic tissue which literally means blood-making tissue. The marrow cavity is also a location for __________________ in adults.

19. Bones also function as a storage site or depot for the element __________________. Not only is calcium important for the __________________ of our bones, calcium plays a major role in our __________________.

20. There are three types of muscle in the body. There is __________________which is found in our internal organs like our stomach, intestines, bladder and respiratory tract. Then there is __________________ which is specifically made for our heart. Finally, there is __________________ which enables us to move.

21. Skeletal muscle is also referred to as __________________ muscle.

22. There are components of skeletal muscle cells made up of what are known as __________________ proteins or filaments. There are two types of contractile proteins: __________________ and __________________.

23. The attachment of muscles to bones is made through tough connective tissue material known as __________________.

24. Connective tissue structures which connect bones to bones are known as __________________.

25. Muscles can only actively __________________.

26. The end of the muscle which is nearer to the main trunk (chest and abdomen) of the body is

known as the _________________ of the muscle. The opposite end of the muscle which is farther from the main trunk of the body is known as the _________________ of the muscle.

27. Bending of a joint is referred to as _________________ while straightening a joint is known as _________________ the joint.

28. Inward movement is referred to as _________________ and outward motion is referred to as _________________.

29. Cells of the nervous systems are called _________________ and consist of a central nucleus-containing area known as the _________________. A single _________________ is present with multiple _________________.

30. An action potential (stimulus) travels into the _________________ "end" and out through the _________________ "end" of a neuron.

31. Axons terminate on _________________ (muscle cells) where _________________ are released to activate muscle contraction. The most common neurotransmitter is _________________ (ACH).

32. Neurons which carry messages out to muscles are known as _________________ or _________________ neurons.

33. Neurons which carry messages inward from sensory organs such as the skin or eyes are called _________________ or _________________ neurons.

34. Conduction of a stimulus is the result of _________________ and _________________ being pumped in and out of the neuron. The element _________________ mediates this process.

35. _________________ interferes with transmission of nerve impulses or neurotransmitter function which results in the desired control of pain and/or movement.

Name ____Date______________________

Lesson 21 Practice Page 2

Use the clues on the next page to solve this crossword puzzle regarding the skeletal, muscular and nervous systems.

Clues:

Across

1. bones of arms and legs

5. making joint angle less

6. type of osteocyte which breaks down bone

9. type of muscle which moves bones

11. "end" of neuron where stimulus departs

12. type of osteocytes which builds bone

16. ends of long bones

17. hip joint

18. type of neuron which results in muscle action

19. part of bone between epiphyses

20. type of joint between long bones

22. making joint angle greater

23. contractile proteins of muscles

24. only intentional action of a muscle

25. bones of skull and pelvis

Down

2. type of joint between flat bones

3. "end" of neuron where stimulus leaves, multiple

4. elbow joint

7. location for red blood cell production

8. nerve cell

9. type of muscle of internal organs other than heart

10. element stored in bones

13. type of neuron which carries sensory information back to brain for interpretation

14. chemicals which move from end of nerve to muscle to trigger contraction

15. lubricant of joints

21. study of parts of body

Name ____Date_______________________

Lesson 22 Practice Page 1

Please fill in the blank(s) for each statement below. Use your lesson to help you.

1. The primary function of the circulatory system is to deliver ________________ and _____________ to all cells of the body.

2. ________________________________ are the smallest blood vessels, are only ________ cell layer in thickness and have windows or openings known as ____________________ to allow certain substances to pass.

3. The ______________ of the fenestrations of capillaries varies throughout the body with some areas being very tightly controlled and other areas like in the _____________________ where whole blood cells freely move in and out of the capillaries.

4. Capillaries are "fed" by _________________________ which in turn receive blood from ____________________ which get blood from the largest artery of all which is the __________.

5. In the human heart there are four ________________________ . The two which receive blood from outside the heart are known as the _____________________ while the two which pump blood out of the heart are known as the __________________________.

6. Blood from the body that enters the heart is __________ in oxygen content and _______ in carbon dioxide. The chamber of the heart it enters is the ________________ atrium.

7. From this chamber, upon contraction of the right atrium, blood moves down through the ____________________________ valve into the __________________ ventricle.

8. Upon contraction of the right ventricle, blood moves out of the heart destined for the _______ to get __________________ and leave behind ____________________.

9. Blood returning from the lungs is ________________ in oxygen and ____________ in carbon dioxide. It enters the heart in the _________________ atrium.

10. From the left atrium, blood moves downward through the ________________________ valve into the _______________ ventricle.

11. Upon contraction, blood leaves the left ventricle passing through the __________________ valve heading to the ________________ through the largest artery known as the __________________.

12. The right and left atrioventricular valves consist of flap-like structures known as the _________ of the valves. Preventing these structures from prolapsing upward into the atria above are tough string-like structures known as _________________________________.

13. If a valve does not close properly which allows blood to "leak" in an unintended direction or is too "tight" (which is known as ________________,) one can hear abnormal sounds known as ________________________.

14. Air moves down from the nose and mouth through the ______________________ which branches into _________________________ which branch further into _________________________.

15. The bronchioles end in small grape-like clusters of air sacs known as ______________________.
_____________________________ wrap around the alveoli to pick up oxygen and leave behind _____________________________.

16. In the body, capillaries returning from cells deliver blood into the smallest veins known as _________________________.

17. Unlike arteries, veins have _________________ walls and have small _______________ within them to prevent backflow of blood.

18. The watery portion of blood is known as ____________________.

19. Red blood cells are _____________________________ while the scientific name for the group of white blood cells are the _______________________________.

20. Red blood cells are produced in the ________________________________ and have the primary job of ____ through the work of a special protein found within them known as _____________________________________.

21. The decrease in the functioning number of red blood cells is known as _______________.

22. _________________ function as part of our body's defense system against invading organisms.

23. ______________________are white blood cells which create antibodies or _________________ which work as ____________________ for later attack of invaders.

24. ___________________, ________________ and _________________ work to mediate inflammation in the body.

25. Carbon monoxide causes problems in our body because it replaces _____________ being carried by _________________________ within red blood cells.

Name ____ Date______________________

Lesson 22 Practice Page 2

Use the clues below to solve this puzzle.

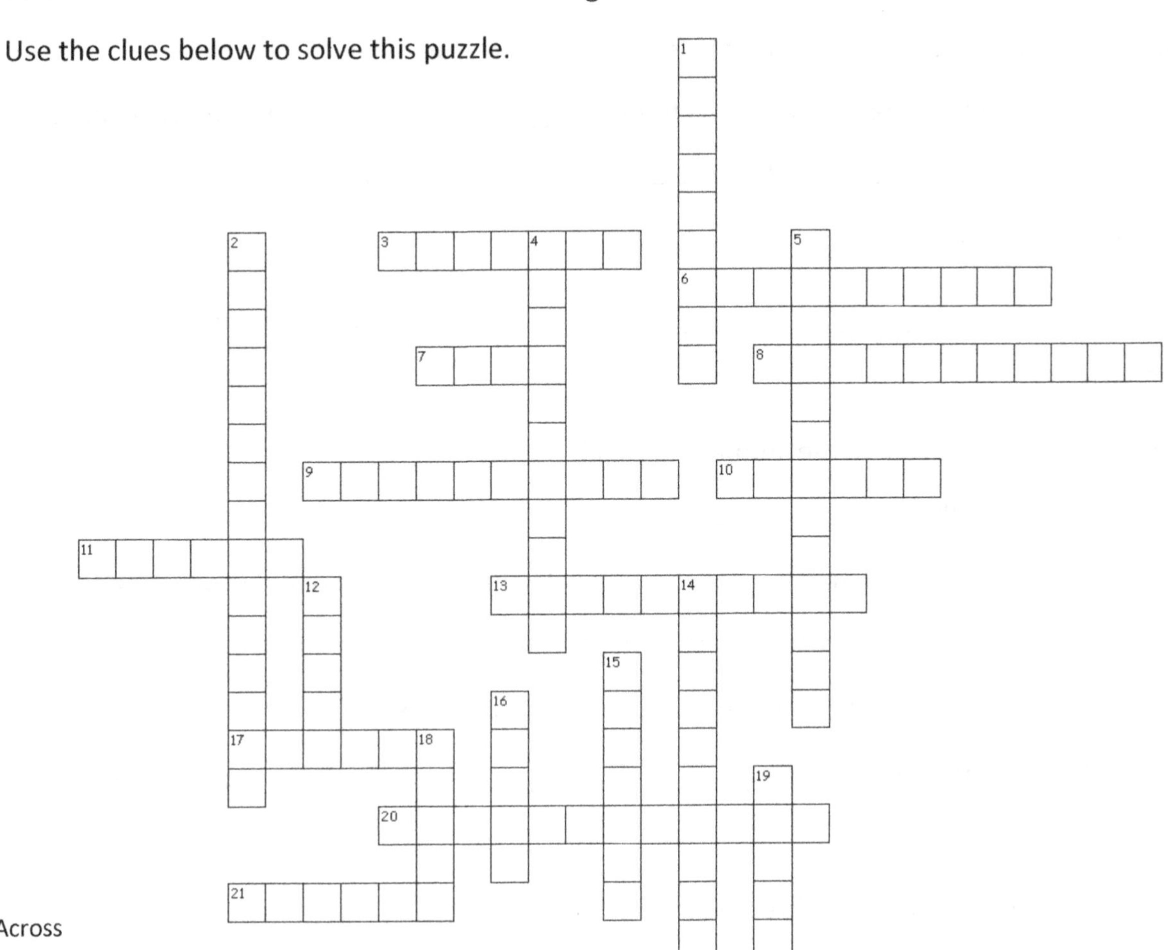

Across

3. tiny veins

6. smaller than an artery but larger than a capillary

7. flap portion of valve

8. most abundant leukocyte

9. chambers of heart which pump blood out of heart

10. location of red blood cell production

11. necessary substance for respiration

13. oxygen-carrying protein of red blood cells

17. low number of erythrocytes

20. red blood cells

21. watery portion of blood

Down

1. smallest blood vessel

2. "strings" of valve which prevents valvular prolapse

4. white blood cells which makes antibodies

5. windows of capillaries

12. "gate" within heart which allows for one direction flow

14. white blood cells

15. fuel of body carried by blood

16. largest artery of all

18. chambers of heart which welcome blood

19. pumper of blood

Name ____Date______________________

Lesson 23 Practice Page 1

Please fill in the blank(s) for each statement below. Use your lesson pages to help you.

1. The main function of the digestive system is to take ___________________ and convert it into glucose to be delivered to _____________ by the ____________________ system.

2. A bite of food is known as a _________________________.

3. Saliva, which contains ______________________________, is mixed with the bolus to begin the digestion process.

4. Food moves from the mouth to the stomach through the ____________________ by the action of ____________________muscles.

5. The stomach mixes the food with substances produced by the chief and parietal cells. The chief cells create ____________________ which reacts with hydrochloric acid from the ____________________ to produce pepsin, a ____________________ enzyme.

6. The "front door" of the stomach is the ____________________while the "back door" of the stomach known as ____________________.

7. The small intestine is divided into three segments: the ____________________ which is closely associated with the pancreas, the ____________________ where microvilli allow for great surface area for nutrient absorption and the ____________________.

8. The pancreas secretes the hormone ____________________ which regulates the ability of glucose to enter ____________________ of the body.

9. The ____________________ is known to have close to 500 specific jobs in the body which includes the ____________________ arriving from the small intestine as well as the manufacture of ____________________ and ____________________ (the storage form of glucose.)

10. At the junction of the ileum and the large intestine (colon) is the ___________ or ______________.

11. The primary role of the large intestine is ____________________ and the production and absorption of ______________which is vital for blood ____________________.

Name ____ Date ______________________

Lesson 23 Practice Page 2

Use the clues below to solve this crossword puzzle about the digestive system.

Across

2. this sphincter is the back door

6. process of taking food and converting to glucose

8. bite of food

9. pancreas secretes this hormone which allows glucose to enter cells

10. these are found in saliva to begin process of digestion

11. colon reabsorbs this

13. first section of small intestine with pancreas as a pal 14. allow for great surface area within small intestine

16. fancy name for spit

Down

1. enzymes which lyse proteins

2. cells of stomach which produce hydrochloric acid

3. cells of stomach which produce pepsinogen

4. tube from mouth to stomach, made of smooth muscle

5. longest section of small intestine

7. large intestine produces and absorbs this vitamin

12. blind tube portion of intestine

15. this sphincter is the front door

Name ____Date________________________

Lesson 24 Practice Page 1

Please fill in the blank(s) for each statement below. Use your lesson pages to help you.

1. The __________________ and __________________ systems of our body work to remove wastes created by cells as well as cell remnants from the body.

2. The renal system consists of our two __________________ and blood vessels which carry "dirty" blood to and "clean" blood away from the __________________.

3. The filtration unit of the kidney is the __________________.

4. Waste products as well as water can leave through __________________ in the walls of the arterioles within the glomerulus, however, __________________, __________________ and __________________ are returned back to circulation and therefore conserved.

5. The kidneys also maintain __________________ of the blood.

6. The kidneys produce hormones which control __________________and the creation of __________________.

7. The presence of __________________ in the urine indicates possible kidney damage or damage to other parts of the urinary system such as the __________________ or __________________.

8. Glucose in the __________________ indicates it has exceeded the threshold capacity of the kidney.

__________________, produced by the __________________, functions to open "doors" of cell membranes to allow glucose to enter.

9. Lack of insulin can result in __________________ levels of blood glucose, presence of glucose in the __________________ and "starving" __________________ all over the body.

10. Urine leaves the kidneys through the __________________ which drain into the __________________.

11. When full, the bladder is emptied to the exterior of the body through the __________________.

Name ____ Date ______________________

Lesson 24 Practice Page 2

Use the clues below to solve this crossword puzzle about the renal and urinary systems.

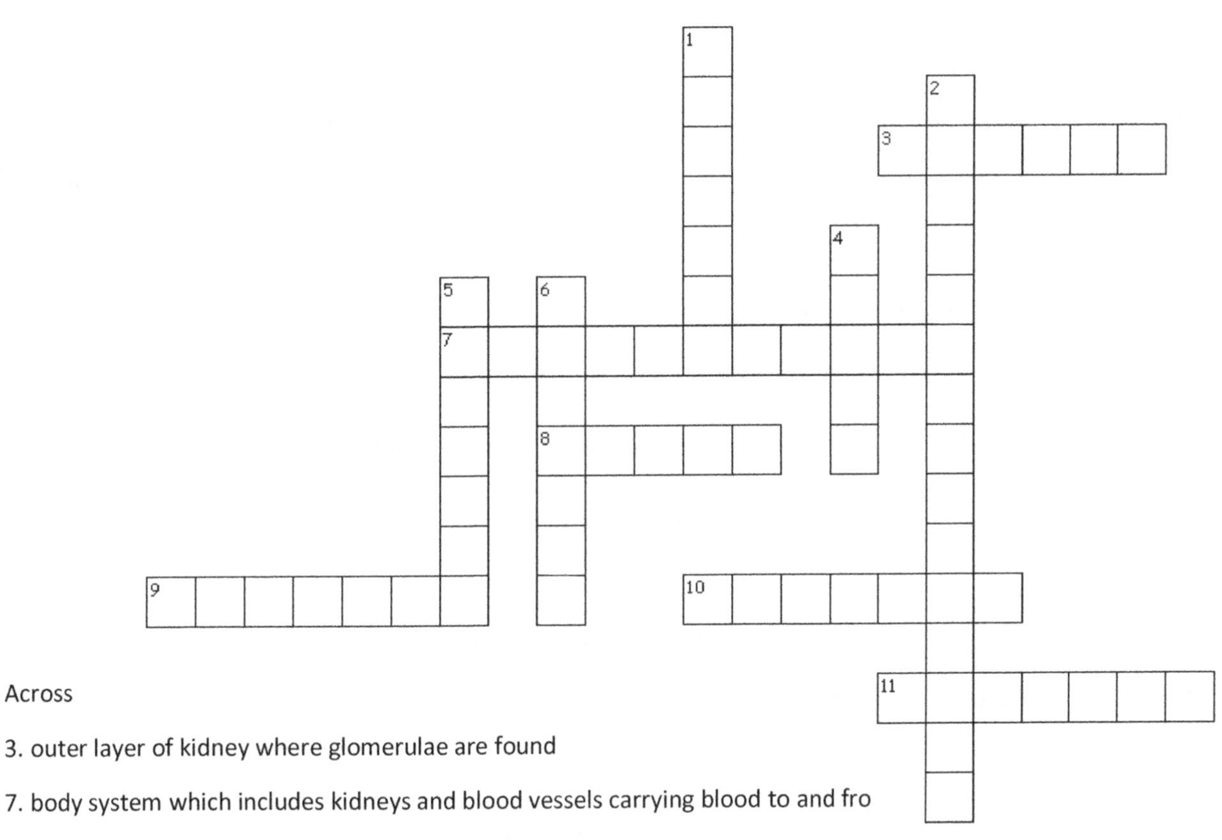

Across

3. outer layer of kidney where glomerulae are found

7. body system which includes kidneys and blood vessels carrying blood to and fro

8. waste-full liquid produced by kidneys

9. inner layer of kidney

10. presence of this in urine means too much is in blood

11. we have two and their job is to filter blood, regulate pH and blood pressure and RBC production

Down

1. these tubes carry urine from kidney to urinary bladder

2. these two elements are also "saved" by the kidneys and not flushed away

4. this gets conserved by kidneys so urine is not too watery

5. this tube empties bladder to exterior of body

6. this hormone produced by pancreas opens "doors" of cells to welcome glucose

Name ____Date______________________

Lesson 25 Practice Page 1

Please fill in the blank(s) for each statement below. Use your lesson pages to help you.

1. The ____________________ system is the body system which regulates the activities of many other body systems through the function of ____________________.

2. There are _____ types of glands: ____________________ which have ducts or tubes in which substances are exported and ____________________ glands which deposit substances directly into the circulatory system.

3. Levels of hormones in the body are maintained using a ____________________ whereby low levels trigger the endocrine gland to begin production of the hormone.

4. The ____________________ is known as the "master" gland as it produces hormones which affect many organs in the body as well as other ____________________ glands which in turn affect body organs.

5. The ____________________ pituitary gland produces thyroid stimulating hormone (TSH), adrenocorticotropic hormone (ACTH), human growth hormone (HGH), prolactin and the gonadotropic hormones.

6. TSH affects rate of ____________________ which is how fast or slow cells utilize ____________.

7. ACTH signals the ____________________to produce anti-inflammatory substances and water conservation hormones.

8. HGH regulates ____________________ in the body.

9. Prolactin regulates development of ____________________ tissues as well as milk production in women.

10. The gonadotropic hormones in women stimulate the ____________________ to produce ____________________ and then maintain pregnancy.

11. The gonadotropic hormones in men stimulate the ____________________ to produce ____________________ and ____________________.

12. Hormones of the posterior lobe of the pituitary gland include ____________________ also known as ____________________ which regulates water conservation by the kidneys and oxytocin which stimulates the ____________________ to contract during birth and milk ____________________ while nursing.

13. The parathyroid glands control ____________________ levels in the body.

14. The ____________________ gland regulates daily sleep and wakefulness patterns.

15. The thymus, while only active in children, functions to produce ____________________ which function throughout life to defend against ____________________.

Name ____Date____________________

Lesson 25 Practice Page 2

Use the clues on the next page to solve this crossword puzzle about endocrine system.

Across

1. this hormone stimulates development of follicles on the ovary in women

4. target organs of ACTH

6. the "master" gland

7. target organs of TSH

9. this hormone stimulates production of testosterone in men

11. this hormone stimulates the uterus to contract during labor as well as milk let-down during nursing

13. this body system utilizes hormones to regulate other body systems

14. target organs of gonadotropins in general

16. "back" side of pituitary gland

17. rate of glucose use by cells

19. this hormone stimulates production of sperm cells in men

20. this tiny gland regulates sleep and wakefulness patterns

21. produced by the posterior pituitary gland to regulate water levels in the body

Down

2. this childhood gland produces killer-T cells for lifelong defense strategies

3. adrenal cortex produces these agents to reduce pain, swelling and redness

5. this hormone stimulates ovulation in women

8. "front" side of pituitary gland

10. HGH which regulates growth is short for ________________

12. this mechanism is based upon low levels triggering action upon producing glands

15. these glands have tubes or ducts to delivery products to body

18. levels of this important element are controlled by the parathyroid glands

Name ____Date______________________

Lesson 26 Practice Page 1

Please fill in the blank(s) for each statement below. Use your lesson pages to help you.

1. Sperm and ova are _________________ cells having under gone meiosis. Sperm are formed in the ____________________ and ova form in the ___________________________.

2. The testis are located outside the body in the ________________. The ovaries are located in the _________________________.

3. Sperm leave the testis through the ____________________________ and continue through the vas deferens to the __________________________.

4. Accessory sex glands which include the seminal vesicles, Cowper's gland and prostate gland provide ___________________and _______________ substances to the sperm to create ________.

5. The ___________________ allows the man to deposit semen into the woman's ___________.

6. The cervix functions like a ____________ to open at specific times in response to ___________ to allow sperm to enter or be closed during pregnancy.

7. The release of the ovum is known as ______________________. The ovum is gathered by the ________________________ and moved down the fallopian tube where _______________________ usually takes place.

8. The embryo moves down to the __________________ to implant and continue development.

9. Blood never mixes directly between the _____________________ and developing ________. Nutrients and wastes along with oxygen and carbon dioxide readily move across _______________ membranes of the ____________________.

10. Pregnancy in humans is __________________ months.

11. _________________________ from the posterior pituitary gland stimulates the uterus to begin contractions while decreasing ___________________ levels allows the cervix to dilate.

12. Should the baby be too large to deliver, a _______________________ may be performed to surgically remove the baby from the uterus.

Name ____ Date ______________________

Lesson 26 Practice Page 2

Use the clues below to solve this crossword puzzle.

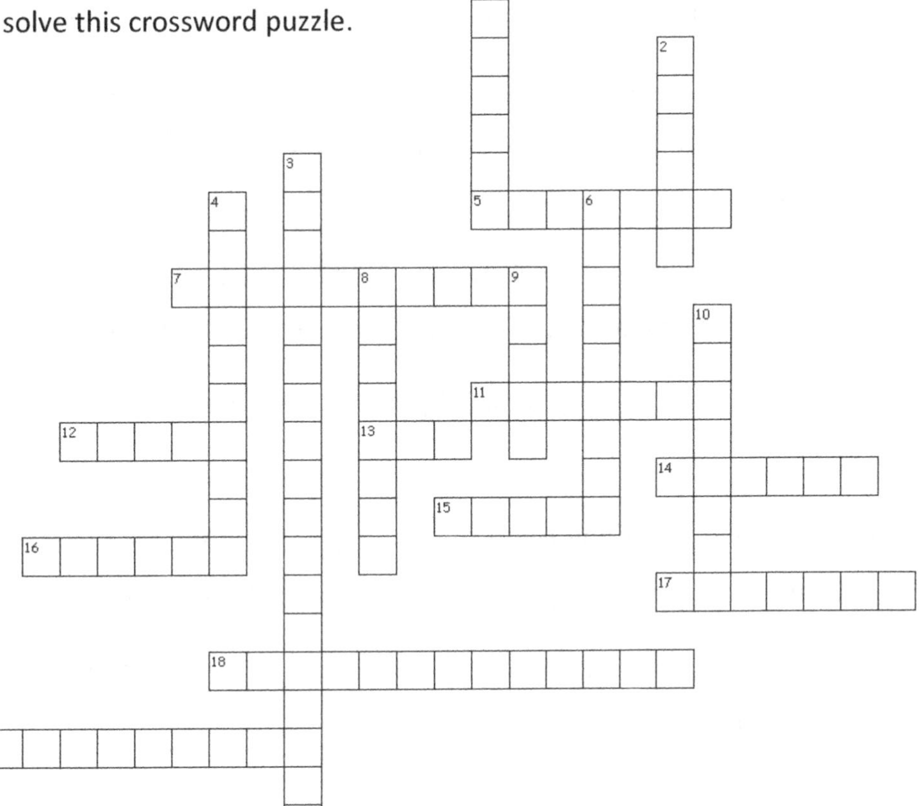

Across

5. process of going from 2N to 1N
7. length of time of pregnancy
11. tube leading from vas deferens to exterior of body
12. location for ovum development
13. female gametes
14. "door" of female reproductive system
15. liquid carrying sperm cells
16. location where embryo implants and grows throughout pregnancy
17. condition of having one-half the total number of chromosomes 1N
18. usual location of fertilization
19. process where gametes join to create new 2N individual

Down

1. location of testis outside body
2. location for sperm and testosterone production
3. includes seminal vesicles, Cowper's gland and prostate gland
4. tube leading from testis to vas deferens
6. process where follicle ruptures to release ovum
8. hormone which causes uterine contractions
9. male gametes
10. location where blood vessels from baby come close to blood vessels of mother but never actually join

Name ____ Date ______________________

Lesson 27 Practice Page 1

Please fill in the blank(s) for each statement below. Use your lesson pages to help you.

1. ______________________________, ________________________, ____________________,

____________________, and ____________________ are our five means of sensing what it happening in our environment.

2. The clear outer covering of our eyes is called the ____________ which transitions to the ____________ which is the white portion on the perimeter of the eye.

3. Tears from the ____________ continually bathe the cornea to maintain hydration and wash away ____________.

4. The colored portion within the eye is the ____________ and creates the circular ____________ which adjusts the amount of ____________ which can enter the eye.

5. The fluid in front of the lens of the eye is the ____________ humor while the fluid behind the lens, which is much thicker, is the ____________ humor.

6. Excess fluid within the eye is ____________.

7. Tiny muscles encircling the lens allow it to change ____________ which in turn allows images to be focused on the ____________.

8. The ____________ is the multilayered surface at the back of the eye which has specialized cells capable of converting light into nerve stimulus.

9. Nerve endings in the retina gather into the ____________ which transmits messages to the ____________.

10. The ears have the capability of converting ____________ energy into ____________ which in turn stimulate nerve endings to allow us to hear.

11. Sound moves into the auditory canal and strikes the ____________ membrane which causes the three tiny bones of the middle ear to vibrate. These vibrations stimulate nerve endings within the fluid-filled ____________ which then send signals by the ____________ nerve to the brain.

12. The semicircular canals within the ear work to maintain _____________.

13. In our nose there are nerve endings which are capable of taking _____________ stimuli created by contact with tiny molecules of substances floating in the air and converting it to nerve impulses.

14. Stimuli within the nose travels to the brain by the _____________ nerve.

15. The main organ of taste sensation is the _____________ .

_____________ on the surface of the tongue house taste buds which consist of taste receptor cells.

16. The primary organ of touch is our _____________. When compared to all other organs in our body, the skin is the _____________ organ and has the most _____________ (weight).

17. The skin consists of _____________ layers: the _____________ which creates squamous cells that form the surface to the skin and the deeper _____________ which houses blood vessels, nerves, glands and fatty tissue.

18. Touch sensations travel by _____________ neurons to the spinal cord and brain.

19. The skin can be divided into regions of sensation. These regions are known as _____________.

Name ____Date______________________

Lesson 27 Practice Page 2

Use the clues below to solve this crossword puzzle.

Across

2. clear covering of the eye
6. largest organ of body enables this sensation
10. three tiny bones of middle ear
14. outer layer of skin with flattened cells
15. fluid-filled snail-shaped structure of inner ear with nerve endings inside
16. curved "onion" which focuses light on retina
17. ear drum
18. sense of flavor gathered by the tongue
19. cranial nerve of scent

Down

1. light sensitive cell layer in rear of eye
3. cranial nerve which connects eyes to brain
4. colorful circle in the eye
5. "job" of the nose
7. three tiny bones work together to enable this sensation
8. hole created by the iris to adjust light entry
9. our eyes enable this ability
11. cranial nerve of hearing
12. clusters of taste buds on surface of tongue
13. deeper layer of skin with blood vessels, nerves, glands and fatty tissues

Name ____Date______________________

Lesson 28 Practice Page 1

Please fill in the blank(s) for each statement below. Use your lesson pages to help you.

1. ________________ is the study of the relationships between living things and their environment.

2. ________________ is defined as being everything, both living and nonliving, which are found in the location where a living organism lives.

3. The ________________ is the segment of the earth where life exists.

4. A ________________ is defined as an area or region of the earth where similar organisms thrive and others do not.

5. The ________________ biome is found at the north and south poles of the earth. The average temperature range is -40 C to -4 C (-40 F to 25 F) with less than 5 inches of precipitation in a year.

6. The temperature range of the ________________ is -26 C to 4 C (-28 F to 39 F) and precipitation is less than 10 inches per year. ________________ and a few mammals are found in this biome.

7. The ________________ biome is characterized by evergreen, cone-bearing trees and has a temperature range of this biome is -10 C to 14 C (14 F to 57 F). The average yearly precipitation is between 12 and 30 inches.

8. The ________________biome is characterized by the predominance of trees which are lose their leaves during the cooler seasons of the year. The average temperature of this biome is 6-28 C (42 F to 82 F) and annual precipitation is 30-50 inches.

9. At the same latitude of the deciduous forest biome is the ________________ biome which experiences temperatures slightly cooler than the deciduous forest biomes and receive only 10-30 inches of precipitation in a year. Abundant grasses and few trees are present in this biome.

10. ________________ biomes are found in regions which are warmer on average than grasslands yet receive less precipitation. The temperature range in desert biomes are 24-34 C (75 F to 94 F) with precipitation being less than 10 inches per year.

11. The rain forest biome can be divided into the ________________ rain forest and the

_______________ rain forest. The tropical rain forest is found all along the equator where the average temperature is 25-27 C (77 F to 81 F). The temperate rain forest is found along the western coast of North America where temperatures range from 10-20 C (50 F to 60 F).

12. The _______________ biome consists of the earth's oceans and seas.

13. The _______________ biome consists of lakes, rivers and streams.

_______________ are where rivers of freshwater mix with the salt water of a sea or ocean.

14. The organisms which are capable of capturing energy from the sun and converting it into useable food for itself (and later, others) are known as _______________. Autotrophs are also called _______________.

15. Heterotrophs "harvest" energy captured from _______________. These organisms are also known as _______________.

16. The consumers which first eats a producer (plant) is known as a _______________ consumer. Organisms which eat a primary consumer are called _______________ consumers.

17. Consumers which have plants as their sole source of food are known as _______________.

18. Consumers which eat both plants and animals (producers and consumers) are known as _______________.

19. _______________ are those organisms which consume dead organisms

20. With each step through an ecosystem away from the producer level, the amount of available energy is _______________.

21. _______________ cycles demonstrate how chemical elements move round-and-round through living (bio-) and then non-living portions (geo-) of the environment.

Name ____________________ Date ____________

Lesson 28 Practice Page 2

Use the clues below to solve this crossword puzzle.

Across

6. eater of road kill

11. biome which receives most rainfall

13. biome characterized by abundant grasses and few trees

14. these are the trees found in the coniferous forest biome

17. you'll only find very short plants and a few mammals living here

18. those who consume autotrophs

19. eater of plants

Down

1. diagram showing how elements move about through the living and non-living segments of the earth

2. everything, both living and non-living, where an organism lives

3. diagram showing how things eat one another

4. trees which lose their leaves in the cool seasons of the year

5. eater of both plants and animals

7. study of relationships of living things and their environment

8. coldest biome of the earth

9. one who consumes producer

10. another name for autotroph

12. eater of animals

15. literally means self-feeder

16. location where some living things thrive and others do not

Made in the USA
Monee, IL
22 April 2024